WOMEN AVIATORS

AMERICAN PROFILES

PROFILES

WOMEN AVIATORS

■

Lisa Yount

☑
Facts On File®

AN INFOBASE HOLDINGS COMPANY

Facts On File, Inc.
460 Park Avenue South
New York, NY 10016

Library of Congress Cataloging-in-Publication Data

Yount, Lisa
   Women aviators / Lisa Yount.
      p. cm. — (American profiles)
      Contents: Katherine Stinson—Bessie Coleman—Amelia Mary
   Earhart—Edna Gardner Whyte—Anne Morrow Lindbergh—
   Jacqueline Cochran—Geraldine Fredritz Mock—Geraldyn Cobb—
   Bonnie Linda Tiburzi—Sally Kristin Ride—Jeana L. Yeager.
      Includes bibliographical references and index.
      ISBN 0-8160-3062-6
      1. Women air pilots—Biography—Juvenile literature. 2. Women in
   aeronautics—History—Juvenile literature. [1. Air pilots.
   2. Women—Biography.] I. Title. II. Series: American profiles
   (Facts On File, Inc.)
   TL539.^68 1995
   629.13′092′2—dc20
   [B]   94-23134

Facts On File books are available at special discounts when purchased in bulk quantities for businesses, associations, institutions or sales promotions. Please contact our Special Sales Department in New York at 212/683-2244 or 800/322-8755.

Text design by Ron Monteleone
Cover design by F. C. Pusterla Design

Printed in the United States of America

IH MP 10 9 8 7 6 5 4 3 2 1

This book is printed on acid-free paper.

To Agnes
in memory of the barnstormer photo

Contents

Introduction

Since the beginning of aviation, men have been trying to keep women out of the air—and failing.

On September 2, 1910, less than seven years after the Wright brothers sent the first airplane into the skies, pioneer aircraft designer Glenn Curtiss was reluctantly giving flying lessons to his first and only woman pupil. His student, Blanche Stuart Scott, had already driven another recent invention, the automobile, across the United States. She was the first woman to do so. Curtiss, however, did not believe that any woman could fly safely. To keep Scott from trying, he had placed a wooden block under her plane's throttle lever. This prevented her from giving the engine enough fuel to lift the craft into the air.

Or so Curtiss thought. On that September morning, he watched in amazement from the ground as Scott's plane suddenly sailed 40 feet through the air. It then landed gently, and its smiling pilot climbed out. "Something must have happened to the throttle block," Scott said demurely.

Blanche Scott was the first woman in the United States to fly a plane, but she was far from the last. Less than a year later, on August 1, 1911, Harriet Quimby became the first American woman to obtain a pilot's license. Soon Katherine Stinson and a few other daring women joined her.

Like all pilots in the early days of aviation, these women risked their lives every time they flew their fragile, unstable contraptions of wood, wire, and cloth. Quimby, for example, was killed in 1912 when she fell out of her plane during an exhibition flight. If a man was killed in an accident like this, people saw it as bad luck. When a woman died, however, it was said to prove that women could not fly. According to the New

York *Sun,* for example, Harriet Quimby's death showed that women "lack strength and the presence of mind and courage to excel as aviators." (Never mind that in the next few years, five-foot-tall, 101-pound Katherine Stinson would not only handle a plane with ease but perform air stunts that few male pilots dared to try.)

Stinson, Bessie Coleman, and most other women pilots before and just after World War I, were stunt or exhibition fliers. There was little else that a pilot, especially a woman pilot, could be. Women were barred from military flying, although they could train military pilots, as Stinson's sister, Marjorie, did. Commercial aviation did not yet exist. Stinson flew mail for a few miles as a stunt in 1913, but when the U.S. Post Office set up its first regular airmail service in 1918, it did not hire women pilots.

The late 1920s and l930s were the golden age of individual achievements in aviation for both men and women. Daring pilots such as Charles Lindbergh, who in 1927 became the first person to fly across the Atlantic Ocean alone, became as popular as movie stars. Amelia Earhart became almost as famous as Lindbergh when she flew across the Atlantic as a mere passenger in 1928. Her fame grew when she made a series of spectacular solo long-distance flights in the 1930s. Anne Morrow Lindbergh came to share her husband's glory when she helped him map air routes over the Arctic and the Atlantic. Record setters such as Jackie Cochran and air-race winners such as Edna Gardner also became "stars."

People in the 1930s were eager to see or hear about women pilots partly because these women's success suggested that flying was safe enough for even "the weaker sex." A transcontinental airline hired Amelia Earhart to persuade potential passengers that commercial flying was safe. Aircraft companies hired women pilots as demonstrators, hoping their customers would believe that if a woman could fly their planes, anyone could.

Of course, allowing women to fly commercial planes on regular runs was another matter entirely. When pilot Ellen

Church applied to Boeing in 1930, the company was happy to employ her—as the country's first flight attendant. One small airline hired Helen Richey as a pilot in 1934, but the all-male pilots' union forced her to quit after a few months.

Women pilots were still barred from combat during World War II. Thanks largely to Jackie Cochran, however, American women did ferry planes from airport to airport, tow targets for antiaircraft gunnery practice, and carry out other noncombat flying jobs. The Women's Airforce Service Pilots (WASPs), which Cochran headed, freed many men pilots for combat duty. The WASPs' service record also helped to destroy myths about women pilots that were still widely believed. WASPs flew huge B-29 Superfortresses, disproving the idea that only men were strong enough to handle heavy planes. WASPs had fewer accidents and lost less time to illness or injury than men with similar training. After studying the WASPs' performance, the air surgeon of the U.S. Army Air Forces said that women were "adapted physically, mentally, and psychologically" to flying high-speed military planes.

Doors opened to women pilots, like many other doors opened to women during the war, slammed shut again when the conflict ended. Homecoming men pilots took all the aviation jobs. One former WASP who applied for positions with a variety of aircraft manufacturers, airlines, and airports said, "I never saw 'no' written in so many different ways." Some women pilots, such as Edna Gardner Whyte, scratched out a living managing small airports or giving flight instruction, but most were no longer in the headlines. One exception was Jackie Cochran, who arranged to borrow a Canadian military jet in 1953 and became the first woman to break the sound barrier.

The 1960s were not much better than the 1950s for women in aviation. When Jerrie Cobb in 1960 took the bold step of trying to become the first American woman in space, she failed. She and 12 other experienced women pilots passed physical and psychological tests given to the country's male astronauts, but the National Aeronautics and Space Administration (NASA) refused to continue the testing or allow the women to undergo astronaut training. Jerrie Mock, who in 1964 achieved Amelia Earhart's goal of becoming the first woman to fly a plane around the world, was more in tune with her times. She described herself as merely a "flying housewife."

American women pilots finally began to see new opportunities in the 1970s. During that decade, the powerful movement for women's rights forced many formerly all-male professions, including both commercial and military aviation, to open their ranks to women. In January 1973, Emily Warner became the first woman to be hired as a pilot by a regularly scheduled U.S. airline. Two months later, Bonnie Tiburzi became the first to be hired by a major airline. Barbara Rainey became the first woman military pilot (for the navy) in 1974. In 1982, Sally Ride fulfilled Jerrie Cobb's dream of becoming the first American woman to go into space.

In the 1980s and early 1990s, women aviators have again dropped out of the headlines—because they are no longer unusual. By 1991, the United States had over 50,000 licensed women pilots. Some 1,200 women flew for commercial airlines, and at least 275, including Bonnie Tiburzi, were captains (the pilot in charge of a flying crew). About 600 women flew for the military, some in combat situations. When a woman pilot such as Jeana Yeager made a major achievement, more attention was given to the achievement than to the fact that the achiever was a woman.

This is not to say that women are now completely equal in the sky. Changes in laws and social attitudes have removed the obvious discrimination that earlier women pilots faced, but more subtle barriers linger. Rosemary Mariner, a navy pilot, says, "I, and those of my generation, will always be foreigners" to the culture of male military pilots. Similarly, airline captain Denise Blankenship says that an airliner's pilot cabin, or cockpit, is "still a male-dominated world."

Other barriers may lie in the minds of women themselves. When asked why American Airlines, for example, has only 200 women among its 9,300 pilots, Bonnie Tiburzi says that only a small number of qualified women have applied for pilots' jobs.

Introduction

Women who feel no barriers, however, continue to be drawn to aviation. In September 1993, for example, 11-year-old Victoria Van Meter of Meadville, Pennsylvania, became the youngest girl to fly a plane across the United States. "I really don't think anything's scary about it," she said. In June 1994 she went on to fly a plane across the Atlantic Ocean, the same journey Amelia Earhart made.

Women in aviation have faced more obvious and longer-lasting discrimination than women in more gender-neutral careers. Women scientists, for example, were at work in many American universities and corporations by 1970, although such women were not numerous and not always fairly treated. Bonnie Tiburzi, however, was still being told she could not be an airline pilot because women airline pilots did not exist.

To overcome such strenuous opposition, the best women pilots have had to develop an extraordinary degree of self-confidence and persistence. "I've learned to be . . . tough. I've *had* to be!" Edna Whyte insisted in her autobiography. In Whyte's opinion, "A shrinking violet would not last a day in a career in aviation. If the men didn't browbeat her to death, the other women would." Similarly, Bonnie Tiburzi described the first women airline pilots as "tough-minded and determined . . . strong-fibered women who had rapped on closed doors and forced them to open."

Why do some women want to fly so badly that they will risk their lives in the air and persist in spite of seemingly endless obstacles on the ground? No doubt there are as many answers as there are successful women pilots. Perhaps the simplest (but also one of the most profound) is the one Amelia Earhart used as the title of a book about flying: *For the Fun of It.* Jackie Cochran put it another way by saying that her love of flying was part of her love of adventure.

For some women, flying was a way to show their competitive spirit. "Winning was the fun" to racer Edna Whyte, and Jackie Cochran also loved winning races and breaking records. For the three women in this book who grew up in poverty—

Coleman, Whyte, and Cochran—flying was also a way to escape their background and "be somebody."

Many women pilots, especially those in the early part of the century, found that flying offered a degree of independence and control over their lives that society denied them on the ground. An early pilot named Margery Brown wrote, "Women are seeking freedom. Freedom in the skies! . . . Flying is a symbol of freedom from limitations." Edna Whyte saw flying as a literal way of taking her mother's advice to "rise above" difficulties and restrictions. Louise Thaden said flying made her feel she was "master of my fate."

Finally, women pilots often saw flying as a magical, even religious experience. Jerrie Cobb wrote that on a high-altitude flight, "I felt that I could reach up to the sun, or touch the stars that were hidden in its glow." Bonnie Tiburzi described the sky as "a world above a world—and you are in it . . . looking out and down at miracles." Cobb, Mock, and Cochran explicitly tied their feelings about flight to their equally strong religious feelings, and some other women pilots have done the same. "Halfway between the earth and sky one seems to be closer to God," Margery Brown wrote.

Perhaps, in the long run, there are no words to describe what makes some women go as far, as fast, and as high as they can in the air. Amelia Earhart wrote to her husband just before she departed on her last flight, "I want to do it because I want to do it. Women must try to do things as men have tried. When they fail, their failure must be but a challenge to others."

Katherine Stinson
(1891–1977)

*Katherine Stinson performed daring stunts and set
records in the early days of aviation.*
(Courtesy National Air and Space Museum, Smithsonian
Institution, SI Neg No. 33444F)

As watchers gasped below, the night sky over Los Angeles
suddenly seemed to burst into flame. The blazing letters "CAL,"
for California, appeared in the air as though some invisible god
were tracing them with a fiery pen.

1

The "pen" was actually a set of magnesium flares attached to the cloth wings of a fragile plane. The writer was not a god but a young woman only five feet tall. She was the first woman pilot to do skywriting at night, a dangerous trick because the flares could easily set the plane's wings on fire. Katherine Stinson continued her exhibition on that evening of December 17, 1915, by flying in loops, turning her plane upside down, and spiraling down to within a scant 100 feet of the ground before she leveled off the plane.

This was only one of the spectacular shows with which Katherine Stinson amazed the people of the United States—and Canada, Japan, and China as well. During her career, Stinson set distance records, invented terrifying air stunts, trained pilots, and became the first woman to carry mail on a plane. Along with her sister, Marjorie, she was a leader among the women who flew in the earliest days of aircraft.

Katherine Stinson was born in Alabama on February 14, 1891. Her parents, Edward and Emma Stinson, had a second daughter, Marjorie, in 1896. Two brothers, Eddie and Jack, made up the rest of what came to be called "the Flying Stinsons." All four would have careers in the air.

As a teenager, Katherine planned to be a piano teacher. She had won a piano in a contest and wanted to take music lessons in Europe. Such lessons would be costly, though, and she knew she would have to earn the money for them herself.

While visiting friends in Kansas City in 1911, Katherine got a chance to ride in a hot-air balloon. By the time she came down, she had decided to learn to fly. Balloons, which people had been flying for more than a hundred years, did not interest her. Those newly invented contraptions called airplanes did. The Wright brothers had flown the first plane just eight years before, and in 1911, there were only 200 licensed airplane pilots in the world. Very few were women. Katherine learned that a pilot could make up to $1,000 a day taking people into the air for short rides. Surely, she thought, flying would be a good way to finance her piano lessons.

But flying lessons also cost money. To pay for them, Katherine sold her piano for $200. After a lot of talking, she persuaded her father to give her the rest of the money she needed. (Her mother was already on her side. "When I began to talk about flying, she already had confidence in me," Katherine said in an interview many years later. "My mother never warned me not to do this or that for fear of being hurt. Of course I got hurt," she added, "but I was never afraid.")

In May 1912, after being turned down by several pilots, Stinson began flying lessons with a famous Swedish pilot, Maximilian Liljestrand. Everyone called him Max Lillie. She made her first flight alone on July 13, after only four hours and 10 minutes of practice in the air. Three days after this solo flight, she received her pilot's license from the Fédération Aéronautique Internationale (FAI). This international air organization, which had its headquarters in France, approved all pilots' licenses at the time.

Stinson was the fourth American woman to receive a pilot's license. At that moment, she was the only one who was flying. The first two, Julia Clark and Harriet Quimby, had been killed in crashes just the month before. After hearing of their deaths, the third pilot, Mathilde Moisant, had abruptly decided to retire.

Stinson had given up all thoughts of a musical career by this time. Flying had become an end in itself. A month after getting her pilot's license, she bought her first plane. She startled the men at the local airfield by cleaning every inch of her new purchase with a scrub brush and soapy water. All through her career, Stinson would insist that her planes be clean and perfectly maintained. Many air accidents, she believed, happened because pilots did not take good care of their aircraft. "It's all right if your automobile goes wrong while you are driving it," she wrote in *American Magazine* in 1917. "You can get out in the road and tinker with it. But if your airplane breaks down, you can't sit on a convenient cloud and tinker with *that!*"

Stinson started giving air exhibitions in July 1913. Because she looked even younger than her 22 years, she was advertised as the "Flying Schoolgirl." In September, while performing for the state fair in Helena, Montana, she became the country's first woman airmail pilot. (Mail had first been carried by air in the United States two years before.) The postmaster in Helena asked the Post Office Department in Washington, D.C., to establish an airmail route from the fairgrounds to downtown Helena. Stinson was sworn in as an official airmail pilot and flew the route for four days, carrying a total of 1,333 postcards and letters. This exhibition performance was typical of airmail flights at the time; the country had no regular airmail service.

The younger members of the Stinson family now demanded to get into the air act. Katherine refused to teach them to fly because she was afraid they would be hurt. So little sister Marjorie—everyone called her Madge—went instead to the Wright school in Dayton, Ohio. In the summer of 1914, she became the ninth American woman to get a pilot's license. She was also the youngest; she was just 18 at the time.

World War I broke out in Europe in August 1914. The United States did not enter the war at that time, but many Americans suspected that the country would be involved sooner or later. Planes began to be used in the war, mostly to observe enemy troop movements from the air. A growing number of young men wanted to become pilots, but the U.S. armed forces had almost no flying schools. Canadian men also wanted to become pilots in order to help Canada's mother country, Great Britain. Canada had no air force, so many Canadians came to the United States for flight training.

The Stinson family opened a flying school at a small airfield in Texas. Marjorie was the school's chief teacher. Her pupils called her the "flying schoolma'am." Emma Stinson handled the business end of the operation. The Stinson School of Flying graduated its first class of five, all Canadians, in November 1915. It continued its operation until August 1917, when the U.S. government forbade all civilian flying. By then the school had trained over 80 pilots.

Katherine Stinson's sister, Marjorie, trained pilots during World War I and was called "The Flying Schoolma'am."
(Courtesy National Air and Space Museum, Smithsonian Institution, SI Neg No. A53203)

Katherine Stinson earned money for the school by performing in air shows around the United States and Canada. In her exhibitions, she pioneered some hair-raising stunts. Some air history accounts say she was the first woman to "loop the loop." (Others claim that this honor belongs to another American woman pilot of the time, Ruth Law.) Not even many men would try this trick. The plane had to fly almost straight up, hang upside down for a moment at the top of the loop, then circle back down to complete the loop. During the climb or the moment of "hanging" at the top, the plane's engine sometimes stopped. If it could not be started again quickly, the plane would crash. One account says that in November 1915 Stinson did 80 loops in succession.

But looping the loop, even 80 times, was not enough for Katherine Stinson. On November 21, 1915, she first demonstrated her own variation of the trick. She called it the "dippy twist loop." In this version, the plane rolled wing over wing at the top of the loop.

In December 1916, Stinson began a tour of Japan and China. No woman pilot had visited these countries before. Some 25,000 people watched Stinson's plane trace a blazing *S* in the night sky over Tokyo on December 15. "Stinson clubs" were formed in the cities where she flew. One Japanese schoolboy wrote to her, "Last night when I saw you . . . flying high up in the darkest sky I could not help to cry: you are indeed Air Queen!"

In China, Stinson gave a private performance of looping and skywriting for the country's leaders. As she was descending, the rudder lever at her feet, which steered the plane, snapped off. She could still guide the plane by moving the stub of the lever. To do so, however, she had to bend over so far that she could not see out. Stinson had to bob up and down, alternating between steering and seeing, until she managed to land. China's president, Li Yung Hung, was so impressed that he called her the "Granddaughter of Heaven." He gave her a silver cup and a diamond pin.

Stinson left Asia on May 6. She may have cut her tour short because she learned that the United States had declared war on Germany in April. As soon as she was home, she tried to join the war as a flier. She was told, however, that no women, pilots or otherwise, were allowed in combat.

Determined to help the war effort, Stinson offered to make a flying tour to raise money for the war relief activities of the Red Cross. The tour would go from Buffalo, New York, to

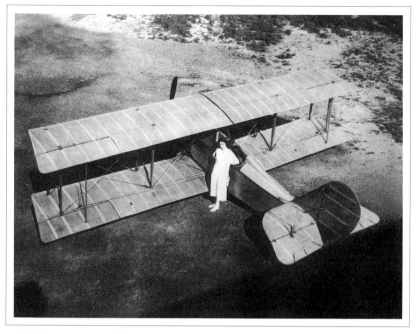

Katherine Stinson flew a Curtiss JN-4 "Jenny" biplane from Buffalo, New York, to Washington, D.C., to raise money for the Red Cross.
(Courtesy National Air and Space Museum, Smithsonian Institution, SI Neg No. A48410)

Washington, D.C., Stinson would drop leaflets and collect donation pledges at every stop.

During her Red Cross tour, Stinson flew a Curtiss JN-4. This sturdy biplane (plane with two sets of wings), known affectionately as the "Jenny," was widely used to train pilots during World War I. (After the war many civilian pilots, including Bessie Coleman, flew Jennies as well.) The Jenny was larger than any plane Stinson had flown before. Nonetheless, she fearlessly took off in it after only five minutes of instruction.

Stinson's Red Cross tour started on June 24, 1917. During the tour she "raced" the Empire State Express train and beat it into Albany, New York, by 34 minutes. She found her way to Washington by using a map pulled from a railroad timetable.

At the end of the trip, Stinson gave Treasury Secretary William McAdoo $2 million in donation pledges.

Stinson then decided to try to set a new air record by flying nonstop from San Diego, California, to San Francisco. She left San Diego at 7:31 A.M. on December 11, 1917. She flew north through heavy fog over Los Angeles, then headed toward the Tehachapi Mountains. She used the Santa Fe railroad tracks as a navigation guide, except for a nerve-racking few minutes when they disappeared into a long tunnel. She had to climb to 9,000 feet, higher than she had ever flown, to cross the mountains at the Tehachapi Pass.

Once she was safely across the mountains, Stinson saw "the beautiful California landscape spread under me like a huge painting" as she flew along at 62 miles an hour. She guided herself by a map that was mounted on rollers, so she could see a great length of it at once. "Towns, cities, farms, hills and mountains passed rapidly," she wrote later. "I never had any fear. The main thing was speed."

At last, Stinson circled over the Golden Gate, the famous entrance to San Francisco Bay. She landed at San Francisco's military fort, the Presidio, at 4:41 P.M. with just two gallons of gas left in her plane's tanks. "Tears came to my eyes as I heard the cheers of thousands of soldiers down below," she wrote. "They were lined up in two files and I landed between them. They rushed up and helped me out of my plane and I was mighty proud." Blaring horns from ships on the city's waterfront added to Stinson's noisy welcome.

Stinson had a right to be proud. She had flown 610 miles nonstop in 9 hours and 10 minutes, a new record for distance and time in the air. The previous nonstop distance record, 510 miles, had been set in November 1916 by Stinson's friendly rival, Ruth Law. "I'll bet Ruth Law is glad a girl and not a man broke her record," Stinson said.

Stinson was still determined to take part in the war overseas. She went to Europe in 1918 and drove ambulances as a volunteer in France for several months. When she returned home after the war's end in November 1918, she did not go back to her flying career. She was exhausted and had caught tuberculosis, a serious lung disease. She had to fight this disease for six years before regaining her health.

Stinson moved to Santa Fe, New Mexico, where the dry air was supposed to help tuberculosis victims. In 1928, she married a Santa Fe lawyer, Miguel Otero. Otero had been a pilot in World War I and shared her love of flying. She also started a new career as an architect. She won several prizes for her home designs.

Stinson won other prizes from those who remembered her as a pioneer of flight. For example, in 1940 the General Federation of Women's Clubs named her one of America's 52 most outstanding women. The group also said she was one of the three living women who had done the most for aviation. Katherine Stinson Otero died in 1977, one year before Sally Ride became an astronaut.

Chronology

February 14, 1891	Katherine Stinson born in Alabama
May 1912	begins flying lessons
July 16, 1912	becomes fourth American woman to get pilot's license
July 1913	begins giving air exhibitions
September 1913	becomes first American woman airmail pilot
November 1915	Stinson School of Flying graduates first class of pilots
November 21, 1915	Stinson first performs "dippy twist loop"
December 17, 1915	becomes first woman to do night skywriting
December 1916– May 1917	tours China and Japan
June 24, 1917	begins flying tour to raise money for Red Cross
August 1917	Stinson School of Flying closes
December 11, 1917	Stinson sets time and distance record by flying nonstop from San Diego to San Francisco
1918	drives ambulances in France; becomes ill and gives up flying career
1928	marries Miguel Otero
1977	Katherine Stinson dies

Further Reading

Adams, Jean, and Margaret Kimball. *Heroines of the Sky.* Garden City, N.Y.: Junior Literary Guild and Doubleday, Doran & Co., 1942. For young adults. Includes a chapter on Stinson. Written in somewhat dramatized style.

Lomax, Judy. *Women of the Air.* New York: Ballantine/Ivy Books, 1988. Contains information on Stinson.

Moolman, Valerie. *Women Aloft.* New York: Time-Life Books, 1981. Contains information on Stinson, including a photo feature on her tour of China and Japan.

Oakes, Claudia M. *United States Women in Aviation Through World War I.* Washington, D.C.: Smithsonian Institution Press, 1978. Contains information on and photographs of the Stinson sisters.

Rogers, Mary Beth, Sherry A. Smith, and Janelle D. Scott. *We Can Fly: Stories of Katherine Stinson and Other Gutsy Texas Women.* Austin: Texas Foundation for Women's Resources and Ellen C. Temple, 1983. Contains a chapter on Stinson.

Underwood, John. *The Stinsons.* Glendale, Calif.: Heritage Press, 1976. First two chapters describe careers of Katherine and Marjorie Stinson. Remainder of book describes the Stinson Aircraft Corporation, which was founded by their brother, Eddie. It pictures many of the planes he designed.

Bessie Coleman
(1892–1926)

Bessie Coleman was the first person of African descent in the world to earn a pilot's license.
(Courtesy National Air and Space Museum, Smithsonian Institution, SI Neg No. 92-8943)

"**B**rave Bessie" Coleman had risked death often during her career as an air show performer. She had pulled her World War I surplus "Jenny" biplane out of spins after its engine died in midair. She had ended dives a mere 200 feet above the earth, zooming over the heads of startled crowds. But just before a show in Waxahachie, Texas, Coleman—the first person of African descent in the world to get a pilot's license—faced her biggest challenge on the ground.

By that time, the fall of 1925, Coleman was nationally known. Both African Americans and whites came by the thousands to see her. And there lay the problem, because Texas was strictly segregated. African Americans not only had to sit separately from whites in the Waxahachie stadium but were expected to go in through separate entrances.

Coleman knew she could not change segregation. That day, however, she spoke up for her fellow African Americans. "I wasn't going to let them humiliate *my* people, who were coming to see me," she said later. "I told them [the stadium managers] I would not fly until they let the blacks through the same gate as the whites." Rather than lose their star attraction, the managers gave in. African Americans still had to sit separately in the stadium, but everyone entered through the same gate.

Bessie Coleman was born on January 26, 1892, in a dirt-floored cabin in Atlanta, Texas. She was one of 13 children. Her family moved to Waxahachie, a small town near Dallas, in 1894. There her father, George, bought a small plot of land and built a house.

George Coleman was almost a full-blooded Indian. In 1901, when Bessie was nine, he moved to Oklahoma, which was then called Indian Territory. He felt he would be better treated there than he had been in Texas. Susan, Bessie's mother, refused to go with him and was left to raise her last four daughters alone. Bessie, the oldest, took care of the younger girls while Susan worked as a cook and housekeeper for a white family.

"My mother's words always gave me the strength to over-come obstacles," Bessie Coleman said later. Susan Coleman could not read or write, but she was determined that her children would "be somebody." She borrowed books from a library wagon that went through the community once or twice a year. Bessie loved these books. "I found a brand new world in the written word," she said. "I couldn't get enough." The Bible and stories about African Americans such as Harriet Tubman were her favorites. She read them aloud to the rest of her family.

Bessie's school shut down during cotton-picking season—from late summer through November or December—while everyone, including the children, worked in the fields. Bessie, who was skilled in math as well as reading, recorded the weight of each sack of cotton her family picked. She made sure the foreman did not cheat them at payment time.

Bessie finished high school, a rare thing for an African-American girl in those days, and vowed to go on to college. She washed and ironed to earn tuition money. In 1910, she enrolled in Langston Industrial College (now Langston University) in Oklahoma, but she had enough money for only one semester. After that she had to return to Waxahachie.

After working and saving for several more years, Bessie Coleman moved to Chicago in 1915. At first she stayed with Walter, one of her older brothers. She worked as a manicurist, caring for the fingernails of customers at the White Sox Barber Shop. The barber shop was owned by the trainer of Chicago's famous baseball club.

On January 30, 1917, Coleman married Claude Glenn, a friend of her brother's. Glenn was 14 years older than Coleman. The reason for their marriage is a mystery, since they apparently never lived together afterward.

The United States entered World War I, and two of Coleman's brothers went to fight in France. "I guess it was the newspapers reporting on the air war in Europe . . . that got me interested in flying," Coleman recalled later. "All the articles I read finally convinced me I should be up there flying and not just reading about it."

At first that goal seemed impossible. Only a few flying schools in the United States admitted women, and none would accept an African-American woman. But during her search for flight training, Coleman made an influential friend. He was Robert

S. Abbott, founder and editor of the *Chicago Weekly Defender* newspaper. Abbott suggested that Coleman get her pilot training in Europe, where prejudice against women and African Americans was not as great as in the United States.

Coleman studied French in night school. She saved money from a new job, as manager of a chili restaurant, and borrowed more from Abbott and other friends. Then, in November 1920, she went to France. For seven months she studied at the country's most famous flight school, run by the Caudron brothers. Her desire to take to the sky was not dimmed even when she saw another student pilot killed in a plane crash. "It was a terrible shock to my nerves, but I never lost them," she told her sister later.

Coleman won her pilot's license from the Fédération Aéronautique Internationale on June 15, 1921. After taking further lessons, she returned to the United States in September. Front-page stories in the *Defender* and other African-American newspapers welcomed her.

Coleman had "grand dreams," not only for herself but for her people. "I decided blacks should not have to experience the difficulties I had faced, so I decided to open a flying school and teach other black women to fly." She knew this task would not be easy. "If I could have a minimum of my desires, I would have no regrets," she said in a memoir taken down by her sister Elois Patterson.

To start her flying school, Coleman needed money. She planned to earn it in the same way most young pilots did in the early 1920s. She would become a "barnstormer," doing stunt flying in air shows and taking paying passengers up for short rides. Some air shows were scheduled in advance, but others were spur-of-the-moment affairs. A group of barnstormers simply flew into an area, rented an empty field from a local farmer, and waited for an audience to gather. After the show, they might sleep in the farmer's barn or outside under the wings of their planes.

After a second trip to Europe for advanced training, Coleman made her first air show flight at Curtiss Field, near New York

City, on the Labor Day weekend of 1922. Several thousand people attended the exhibition, which was billed as "the first public flight of a black woman in this country." As she was to do in most of her future shows, Coleman flew a "Jenny" (Curtiss JN-4), the same kind of aircraft Katherine Stinson had used during her Red Cross tour. Six weeks later, Coleman staged another exhibition in Chicago at the Checkerboard Airdrome (now Midway Airport). Robert Abbott and the *Chicago Defender* paid for both exhibitions.

"Queen Bess," as the *Defender* called her, amazed the crowds at her air shows with aerobatic stunts as impressive as Katherine Stinson's. They included loops, spirals, and low dives. She always made sure her audience got its money's worth. When a woman she had hired as a parachute jumper lost her nerve, Coleman found another pilot to fly the plane, strapped on the parachute harness, and made the jump herself.

Coleman also became known as "Brave Bessie"—for good reason. In one exhibition, her Jenny's motor died when she was at the top of a loop. Instead of curving over gently to complete the circle, the plane started to dive.

Coleman had to turn the dive into a landing. Landing a Jenny was not easy, since the plane had no brakes. It just had to roll to a stop. Even after she straightened the plane out, Coleman knew she was coming in much faster than she should. Fortunately, that particular field had a long runway. The Jenny finally stopped with just 200 feet of runway left.

The crowd cheered, thinking the whole thing was part of the show. Only Brave Bessie knew how close to disaster she had come. But she kept her sense of humor about it. After the landing, a little boy asked her if her plane was supposed to have stopped in the sky like that. She just laughed and said, "No, it wasn't."

Coleman's luck did not always hold. In an exhibition in southern California in 1923, her Jenny crashed and she suffered several broken ribs and a broken leg. But Brave Bessie sent a telegram from her hospital bed: "Tell them all that as soon as I can walk I'm going to fly!"

Bessie Coleman

Bessie Coleman often wore a military-style uniform while
performing air show stunts.
(Courtesy National Air and Space Museum, Smithsonian Institution,
SI Neg No. 84-14782)

In fact, Coleman was not able to fly again for almost two more years. Her 1925 tour of Texas and the South, however, was the most successful tour she had ever made. It included lectures in theaters and schools as well as exhibitions at air shows. By the end of the year, she wrote her sister that she had almost enough money to open her flying school. First, however, she would make another tour in Florida.

As part of her Florida tour, Coleman planned to perform in an air show for the Negro Welfare League of Jacksonville on May Day of 1926. She had just made the final payment on a new Jenny. A white mechanic, William Wills, flew the plane from Dallas to Jacksonville. He had to make two unscheduled landings on the way because of engine trouble.

Early on the morning of April 30, Coleman and Wills took the Jenny up for a test flight. Wills was in the pilot's seat. Coleman was planning to make a parachute jump during her May Day exhibition, and she wanted to study the area to find a good jump site. She did not fasten her seat belt because she was too short to see out of the plane when she was strapped in. She also did not wear her parachute.

After circling over the racetrack where Coleman would perform, Wills began to fly back to the airfield. Then the plane suddenly speeded up and went into a dive. At 500 feet above the ground it turned upside down, throwing Coleman out. She was killed when she hit the ground. The Jenny crashed into a nearby field, killing Wills as well. A friend who had witnessed the crash distractedly tossed a lighted cigarette onto the gasoline-soaked ground near the wreckage, and the plane immediately went up in flames.

When police examined the remains of the plane, they found a wrench jammed in its control gears. The wrench had locked the controls in the down position, making recovery from a dive impossible. No one ever learned how it got there. Some people later suggested that it might have been put there by someone who did not like to see an African-American woman become so popular, but no evidence was ever found to support this. Most likely Coleman's death was accidental.

Following a memorial service in Jacksonville, Coleman's friends took her body back to Chicago. Over 5,000 people attended her funeral. She was buried in Lincoln Cemetery.

Ten years after Coleman's death, Robert Abbott wrote an editorial in the *Chicago Weekly Defender* that said in part, "Though with the crashing of the plane life ceased for Bessie Coleman, she inspired enough members of her race by her courage to carry on in aviation, and what they accomplish will stand as a memorial to Miss Coleman." His words were fitting. At about this same time, another African-American woman pilot, Willa Brown, established the first African American–owned private flight school that was approved by the U.S. government. Brown's school, the Coffey School of Aeronautics, helped to train the pilots of the 99th Pursuit Squadron. This all–African American fighter squadron won many medals for bravery during World War II.

Coleman's influence continued long after that. In 1975, three African-American women in Gary, Indiana, all wives of pilots, formed a club they called the Bessie Coleman Aviators. The club welcomed women of any race, but especially African-American and other minority women, who were interested in flying. Club activities included a free yearly ground school for new members who wanted to become pilots.

Bessie Coleman also has a prominent place in a mural at Lambert–St. Louis Airport in Missouri that honors African Americans' achievements in aviation. The mural pictures 75 African Americans, including Mae Jemison, the first African-American woman chosen to be an astronaut. In an afterword that appeared in Doris Rich's recent biography of Coleman, Jemison wrote, "To me [Coleman] is that ephemeral daydream of adventure, strength, audacity, and beauty that we all seek, hope, and somehow know must be present in the world."

But the inspiration that Brave Bessie Coleman provided is perhaps best described by a 12-year-old Jacksonville girl. A letter from the girl was found in Coleman's pocket after her fatal crash. "I am writing to you to congratulate you on your brave doings," the letter said. "I want to be an aviatrix [woman aviator] when I get [to be] a woman. I like to see our race do brave things. I am going to be out there to see you jump from the airplane. I want an airplane of my own when I get [to be] a woman."

Chronology

January 26, 1892	Bessie Coleman born in Atlanta, Texas
1901	Susan Coleman left to raise four daughters alone
1910	Bessie Coleman enrolls in Langston Industrial College (Oklahoma)
1915	moves to Chicago
January 30, 1917	marries Claude Glenn
November 1920	goes to Europe for flight training
June 15, 1921	becomes first person of African descent to receive a pilot's license
September 1921	returns to United States
September 1922	gives first air show performance
1923	suffers severe injuries in plane crash
1925	makes successful tour of Texas
April 30, 1926	Coleman dies in plane crash in Jacksonville, Florida

Further Reading

Holden, Henry M. *Ladybirds*. Mt. Freedom, N.J.: Black Hawk Publishing Co., 1991. Several pages on Coleman include interesting quotes from the privately published memoir of Coleman written by her sister Elois Patterson.

King, Anita. "Brave Bessie: First Black Pilot." *Essence,* May and June 1976. Interesting biographical article on Coleman, in two parts.

Rappaport, Doreen. *Living Dangerously: American Women Who Risked Their Lives for Adventure*. New York: Harper-Collins, 1991. For young adults. Includes a chapter on Coleman, in dramatized style.

Rich, Doris L. *Queen Bess: Daredevil Aviator*. Washington, D.C.: Smithsonian Institution Press, 1993. For young adults. Only book-length biography of Coleman. Includes background information about African-American society in her time.

Smith, Jessie Carney, ed. *Epic Lives: One Hundred Black Women Who Made a Difference*. Detroit: Visible Ink Press/Gale Research, 1993. Includes a chapter on Coleman.

"They Take to the Sky." *Ebony,* May 1977. Describes the Bessie Coleman Aviators, a club for women interested in aviation that was founded in Bessie Coleman's honor.

Amelia Mary Earhart
(1897–1937)

Amelia Earhart vanished at sea while attempting to fly around the world in this Lockheed Electra. Even before her mysteriouis disappearance, her achievements in long-distance flying had made her the world's best-known woman pilot.
(Courtesy National Air and Space Museum, Smithsonian Institution, SI Neg No. A45874)

"Sos KKAQQ! SOS KKAQQ!" The voice speaking the three- letter signal of distress and her plane's identifying call letters was weak, but the amateur radio operator in Los Angeles recognized it at once. It belonged to the most famous woman pilot in the world: Amelia Earhart. He may have been the last person to hear it.

By the time the operator heard Earhart's call for help on that morning of July 2, 1937, those closest to her already knew she was in trouble. Traveling with a navigator, Fred Noonan, Earhart was trying to fly her Lockheed Electra around the world. She was late for her landing at Howland Island, a speck of land in the vast South Pacific. If she missed the island, she would run out of gas and crash at sea.

When word spread that Earhart was missing, a huge search was launched. Pilots—including famous pilots and women pilots—had been lost at sea before, but there had never been a response like this. Even so, no trace of Earhart, Noonan, or their plane was ever found.

Earhart's disappearance added an element of mystery to a reputation developed through masterful publicity campaigns, genuine achievements in long-distance flying, and personal charm and courage. These factors combined to make Amelia Earhart the best-known woman pilot in aviation history.

Amelia Mary Earhart was born on July 24, 1897, in the home of her grandmother in Atchison, Kansas. She and her younger sister, Muriel, spent much of their childhood there. Edwin Earhart, their father, was a railroad lawyer who traveled a great deal. It was often simplest for the girls to stay with their wealthy grandmother.

Amelia Otis tried to teach her granddaughters to be ladies, but she was not very successful with her namesake. Young Amelia preferred climbing over fences to going through gates. She loved riding horses as much as reading books. She and Muriel sat for hours in an old carriage in the barn, inventing what Amelia called "imaginary journeys full of fabulous perils." They built a "rolly coaster" with wooden tracks to carry them from the top of a tool shed to the ground. Amelia said riding in it was "just like flying."

After grandmother Otis died in 1912, Amelia and Muriel lived with their parents. The family moved often, sometimes living close to poverty, because Edwin Earhart was an alcoholic and lost several jobs. Amelia seldom had time to make friends with her schoolmates; she went to six different high schools in four years. The yearbook of Hyde Park High School in Chicago, from which she graduated in 1916, called her "A.E.—the girl in brown who walks alone."

Using an inheritance from Grandmother Otis, Amelia went for a year to Ogontz School, a college preparatory school for women in Philadelphia. But on a Christmas visit to her sister in Toronto, Canada, in 1917, she saw several young men who had been wounded in World War I. "Four men on crutches . . . changed the course of existence for me," she said later. She dropped out of Ogontz and went to work as a nurse's aide in Toronto's Spadina Military Hospital. Her experiences there made her a pacifist, a believer that all wars were wrong.

Earhart's feelings about war did not keep her from enjoying the sight of Royal Flying Corps cadets training at a nearby airfield. Once a plane flew low over the spot where Earhart and a friend were standing. The other girl ran away, but Earhart remained. "Mingled fear and pleasure . . . surged over me," she wrote later. "I believe that little red airplane said something to me as it swished by."

Earhart entered Columbia University in New York City as a premedical student in 1919. After a year, however, she moved to southern California, where her parents then lived. She took her first plane ride there in 1920. "As soon as we left the ground, I knew I myself had to fly." She began flying lessons in January 1921 with a woman teacher, Neta Snook. Earhart got her pilot's license on December 15, 1921.

Unlike Katherine Stinson and Bessie Coleman, Earhart did not want to be a barnstormer. She disliked the circuslike atmosphere of air shows. Since there were few other flying jobs in the 1920s, especially for women, she assumed she would fly just "for the fun of it."

By 1926 Earhart, thought she had found her real career: social work. She had moved to Boston with her sister and her mother, Amy, who was now divorced from her father. There Earhart took a job at Denison House, a settlement house that helped poor immigrant families.

Charles Lindbergh made the first solo flight across the Atlantic Ocean on May 20, 1927, and became instantly famous. An

American heiress living in England, Amy Phipps Guest, offered to pay the expenses for a woman to fly the Atlantic. The woman, however, had to be "the right sort of girl": pleasant in appearance and well behaved.

Mrs. Guest asked a friend, George Palmer Putnam, to find the right sort of girl. Putnam's grandfather had founded a large publishing company, G. P. Putnam's Sons, and Putnam worked for the company. It had published *We,* Lindbergh's book about his flight, and the book had become a best seller. Putnam hoped a book written by the first woman to cross the Atlantic in a plane would do equally well.

Through friends of friends, Putnam heard about Amelia Earhart. When they met in April 1928, he saw that Earhart not only met Mrs. Guest's requirements but looked strikingly like Charles Lindbergh. A genius at creating publicity, Putnam began planning to call Earhart "Lady Lindy."

Earhart, for her part, was eager to take part in such a "grand adventure," even though she soon learned that the adventure, for her, would be limited. She would receive no pay. Worse still, she would not be allowed to fly the plane, a Fokker trimotor christened the *Friendship*. Wilmer Stultz, an experienced pilot, would do most of the work. Louis Gordon would assist him as copilot and mechanic. Earhart's only job would be keeping the flight log. Even as a passenger, however, she knew she would be risking her life. Nineteen pilots, including three women, had been lost at sea while trying to cross the Atlantic.

The *Friendship* left Trepassey, Newfoundland, on the morning of June 17, 1928. Stultz had to fly through fog and snow without instruments. Ice formed on the plane's wings, and its added weight forced the craft down toward the water. The plane's radio failed, so it could not get position information from ships passing below. The *Friendship* was almost out of fuel by the time its crew spotted land.

Throughout the difficult flight, Earhart crouched behind the extra fuel tanks. She filled in her logbook and took notes for the book she had promised Putnam she would write. At one point she wrote that she was "getting housemaid's knee kneeling here gulping beauty."

The *Friendship* landed at Burry Port, Wales, after 20 hours and 40 minutes in the air. The crowd that gathered soon became a pushing, shouting mob. It gave Earhart her first, terrifying taste of fame.

The crew went on to London, where Earhart met even larger crowds. She also was introduced to the cream of British society. She described her eight days in England's capital as "a jumble of teas, theaters, speech making, exhibition tennis, polo and Parliament, with hundreds of faces crowded in." Returning to New York by ship, she was greeted by a huge reception that Putnam had set up. Earhart told everyone that all credit for the trip should go to Stultz and Gordon, but "it was evident the accident of sex made me the chief performer in our particular sideshow."

In addition to finishing her book about the flight, Earhart visited more than 30 cities and gave 200 interviews and 100 speeches during the next six months. People flocked to hear her, just as they watch celebrities on television today. Indeed, Earhart's popularity went far beyond anything Putnam had imagined. Americans of the 1920s saw pilots as heroes, and Earhart was perfect for the part. Brave yet modest and charming, she appealed both to the newly independent young women of the time and to their parents.

At first, Earhart hated the publicity "sideshow," but she soon realized she could harness it for her own purposes. In the years to come, she would use her lectures, interviews, and public appearances to promote commercial aviation, pacifism, and rights and jobs for women—especially women pilots. Fees from lectures, product endorsements, and other activities also helped her pay for planes and their upkeep. Famous as she was, Earhart could not forget that she had been merely what she called "baggage" on the *Friendship* flight. She was determined to prove her skill as a pilot. In August 1929, partly for this reason, she flew in the first air race for women. This cross-country race was part of the yearly National Air Races, sponsored by the National Aeronautics Association. Humorist Will Rogers jokingly called it "the Powder Puff Derby," and the name stuck. It was not a lightweight event, however. All 19 contestants had brushes with death during the nine-day race, and one was killed. Sixteen of the determined women nonetheless finished the race. Earhart came in third.

After the race, Earhart spoke to her fellow contestants about forming an association of women pilots. The idea appealed to them, and the organization was set up the next year. At Earhart's suggestion, it was called the Ninety- Nines, after the number of its charter members. In 1931, Earhart became the group's first president.

George Putnam was still managing Earhart's career, but he wanted to do more than that. Six times he asked her to marry him. She finally agreed, but she still had doubts. On the night before their wedding she wrote to him, "You must know again my reluctance to marry, my feeling that I shatter thereby chances in work which means most to me. . . . Please let us not interfere with the other's work or play. . . . I may have to keep some place where I can go to be by myself now and then, for I cannot guarantee to endure at all times the confinement of even an attractive cage." Earhart and Putnam nonetheless married on February 7, 1931, at his mother's home in Noank, Connecticut.

Earhart kept her independence and her maiden name, but she found herself more pleased with married life than she had expected. She once said, "Ours is a reasonable and contented partnership, my husband with his solo jobs and I with mine; but the system of dual control works satisfactorily, and our work and our play is a great deal together."

In Amelia Earhart's mind, only one feat would prove that she was a real pilot: flying alone across the Atlantic, something only Lindbergh had done. Flying a single-engine Lockheed Vega, she set out from Harbor Grace, Newfoundland, on the evening of May 20, 1932, to achieve this goal.

At first Earhart flew through a clear, starry night, but then a storm surrounded her. She had only her instruments to guide her—and one of those, the altimeter, suddenly failed. Without it she could not tell how far above the sea she was.

Earhart tried to fly above the storm, but the cold air at the higher altitude made ice form on the Vega's wings. The weight of the ice sent the plane into a spin, during which it dropped 3,000 feet. Only after the warmer air below melted the ice could Earhart regain control. She then looked out her window and saw foam-tipped waves "too close for comfort."

For the next several hours, Earhart alternated between flying high and gathering ice and flying low and risking hitting the water. Then, when she finally escaped from the storm, she discovered a new difficulty. The pipe that carried exhaust gas from her engine was lit by blue flames that spurted through a

Amelia Earhart startled the people of the Irish countryside when her plane landed in a meadow near Londonderry. They could not believe that she had just become the first woman to fly alone across the Atlantic.
(Courtesy National Air and Space Museum, Smithsonian Institution, SI Neg No. 80-11042)

broken weld. If the fire ate through the pipe, the Vega might explode.

Dawn came at last, almost blinding Earhart. After putting on dark glasses, she saw that the gauge on the Vega's reserve gas tank was broken. She could not tell how much fuel she had. The damaged exhaust pipe was also vibrating badly. She set down the plane on the first safe land she could find, a meadow near Londonderry, Ireland. The landing terrified several cows and startled their herder almost as much.

Earhart had flown the Atlantic in 15 hours and 18 minutes, the fastest crossing up to that time. Predictably, her landing was followed by a "sideshow" even bigger than the one that had greeted the *Friendship*. This time, however, Earhart was prepared for it—and she knew she had earned her fame. Awards for her flight included the cross of the French Legion of Honor, the Harmon Trophy for best woman pilot of the year, and the National Geographic Society's gold medal. On July 29, Congress awarded her the Distinguished Flying Cross. She was the first woman to receive this high honor.

In the mid-1930s, Earhart told a friend, "It's a routine now. . . . I make a record, and then I lecture on it. That's where the money comes from. Until it's time to make another record." She set a speed and distance record in 1932 for nonstop, cross-country flying—2,447 miles (between Los Angeles and Newark, New Jersey) in 19 hours and 5 minutes. Three years later, she beat that record by almost two hours. In 1935, she also became the first person to fly alone across the Pacific Ocean from Hawaii to San Francisco. She made a record-setting flight from Mexico City to Newark, across the dangerous Gulf of Mexico, in that year as well.

In 1935, Earhart became a visiting career counselor for women and advisor in aeronautics at Purdue University in Indiana. Purdue bought her a new plane, a twin-engine, ten-seat Lockheed Electra. The plane was to be a "flying laboratory" for studying human reactions and mechanical performance during long-term and high-altitude flight.

Amelia Earhart

Amelia Earhart with her husband, George Palmer Putnam (left), and Eleanor Roosevelt. Putnam's skill at publicity helped make Earhart famous and allowed her to meet such people as the first lady. Roosevelt was very interested in both aviation and women's achievements. She became friends with Jacqueline Cochran as well as with Earhart.
(Courtesy National Air and Space Museum, Smithsonian Institution, SI Neg No. 82-8681)

For Earhart, that translated to flying around the world. No woman pilot had made such a flight. Furthermore, no one had flown around the world near the equator. Earhart chose this route, which she expected to cover some 29,000 miles.

Putnam, as always, did most of the planning for Earhart's trip. He obtained permits from the governments of countries where Earhart expected to land. He sent fuel and spare parts to airports along her route. Earhart, meanwhile, tried to learn the uses of all the Electra's controls and instruments—over 100 of them.

Earhart began her trip by flying from Oakland, California, to Hawaii on March 17, 1937. As she was leaving Hawaii three days later, however, the Electra tipped over on the runway and was badly damaged. It had to be shipped back to Lockheed's California factory. Even with a repair crew working overtime, the plane was grounded for two months.

The delay forced Earhart to make important changes in her plans. She knew she would have to take another pilot with her as a navigator. The man she had first wanted was no longer free, so she chose another pilot, Fred Noonan. Changes in weather conditions also forced Earhart to reverse her route, flying eastward instead of westward. This meant that the hardest part of the trip, the flight across the Pacific, would come at the end.

Earhart and Noonan took off from Miami, Florida, on June 1. Just before she left, Earhart told a friend, "I have a feeling there is just about one more good flight in my system." The Electra hopped down the eastern coast of Central and South America for six days, then flew across the South Atlantic to Africa. Earhart and Noonan crossed northern Africa, then proceeded to India and Southeast Asia.

On the morning of July 2, Earhart and Noonan left Lae, New Guinea. They had traveled 22,000 miles in 30 days, and both were becoming tired. They now faced the hardest single leg of the trip: 2,556 miles across the Pacific to Howland Island, a dot of land only two miles long and three- quarters of a mile wide. Howland was the only place they could land and refuel before the Electra ran out of gas.

The Coast Guard ship *Itasca* had been stationed at Howland and told to track Earhart by radio. The *Itasca* transmitted a signal that Earhart should have picked up with her automatic direction finder. Earhart, however, had left the long antenna for her direction finder in Miami, so the direction finder was useless. She also needed the antenna to send messages with her telegraph key, which could carry over longer distances than her voice. Earhart had left the antenna behind because she did not think telegraph communication was important.

The *Itasca* began hearing Earhart's voice at 2:45 A.M. on the day after she left Lae. (The International Date Line lay east of Lae, so it was still July 2 on Howland.) She was scheduled to reach Howland at about 6 A.M. Around that time, she repeatedly asked the ship to find her position and send this information back to her, but her signals were too brief for the *Itasca* to be able to do this.

At 7:42, Earhart told the *Itasca,* "We must be on you but cannot see you but gas is running low." Frustrating attempts at communication continued for another hour. Then all signals from Earhart stopped.

On hearing that Earhart was lost, the United States launched the most massive search ever made for a single lost plane. It involved 10 ships, 65 planes, 4,000 men and cost $250,000 a day. For 16 days, the searchers scanned 250,000 square miles of ocean. They found nothing.

Some people saw a mystery in Earhart's disappearance.

During World War II, stories began to circulate that she had been on a spy mission to find out what islands the Japanese controlled in the South Pacific. The stories claimed she had been taken prisoner or even executed by the Japanese.

Debates about Earhart's fate continue to this day. In fact, however, there probably was no mystery. No evidence of the supposed spy mission or capture by the Japanese has survived careful study. Most likely Earhart's plane simply missed Howland, ran out of gas, and sank in the water.

Partly because of her dramatic disappearance, Amelia Earhart has remained far better known than any other woman pilot. Significant as her accomplishments were, they may not merit such a degree of fame. Yet something about Earhart was unique, and almost all who met her seemed to sense it. She helped people see flying as not merely a technical achievement but a symbol of human courage and aspiration. She thus furthered the cause of all women pilots and of aviation itself.

Chronology

July 24, 1897	Amelia Earhart born in Atchison, Kansas
1916	graduates from Hyde Park High School in Chicago
Christmas 1917	decides to become nurse's aide in Toronto
1919	becomes premedical student at Columbia University
1920	takes first plane ride
December 15, 1921	earns pilot's license
1926	moves to Boston; begins social work at Denison House
May 20, 1927	Charles Lindbergh becomes first person to fly alone across the Atlantic
April 1928	Earhart meets George Palmer Putnam, agrees to join flight across the Atlantic
June 17, 1928	becomes first woman to cross the Atlantic by plane
August 1929	takes part in first air race for women
1930	Ninety-Nines founded
1931	Earhart becomes group's first president
February 7, 1931	marries George Palmer Putnam
May 20, 1932	becomes first woman to fly alone across the Atlantic

1935	sets cross-country speed record; becomes first person to fly alone from Hawaii to San Francisco; becomes career counselor and aeronautics advisor for Purdue University
June 1, 1937	leaves with Fred Noonan from Miami, Florida, on round-the-world trip
July 2, 1937	vanishes over the Pacific near Howland Island

Further Reading

Cobblestone, July 1990. For young adults. Whole issue is devoted to articles about Earhart.

Earhart, Amelia. *For the Fun of It.* New York: Brewer, Warren and Putnam, 1932. Earhart describes her own flying, including her solo Atlantic crossing, and adventures of other women in aviation.

Gillespie, Richard. "The Mystery of Amelia Earhart." *Life,* April 1992. One of many articles and books describing attempts to solve mystery of Earhart's disappearance.

Hamill, Pete. "Leather and Pearls: The Cult of Amelia Earhart." *Ms.,* September 1976. Good review of Earhart's career, including theories about her disappearance.

Moolman, Valerie. *Women Aloft.* New York: Time-Life, 1981. Contains extensive information about Earhart.

Morissey, Muriel Earhart. *Amelia, My Courageous Sister.* Santa Clara, Calif.: Osborne Publisher, 1987. Biography of Earhart by her sister.

Nathan, Dorothy. *Women of Courage.* New York: Random House, 1964. For young adults. Contains well-written chapter on Earhart.

Putnam, George Palmer. *Soaring Wings.* Biography of Earhart by her husband. New York: Harcourt, Brace, 1939.

Rich, Doris L. *Amelia Earhart.* Washington, D.C.: Smithsonian Institute, 1989. Recent biography of Earhart.

Wade, Mary Dodson. *Amelia Earhart: Fying for Adventure.* CIP/Millbrook, 1992. For young adults.

Edna Gardner Whyte
(1902–1993)

Edna Gardner Whyte won many air races in the J6-5
Aristocrat and other planes, especially during the 1930s.
(Courtesy the Ninety-Nines, Inc., International Women Pilots
Resource Center, Archives Department,
Oklahoma City, Okla.)

*H*er plane's engine revved and ready to go, Edna Gardner Whyte sat with the other racers entered in the women's free-for-all. In a moment, the starter would lower his flag, signaling the beginning of the race. Most racers watched the flag. Whyte, however, watched the starter's arm. The minute she saw a muscle twitch, she took off. This trick often got her into the air seconds before the other racers.

Whyte knew that seconds counted in a closed-course race, the air equivalent of a sprint. Flying fast and low around the tall posts or pylons that marked the race course was another way to gain seconds, and Whyte was an expert at "polishing the pylons." Some pilots might fly for "the fun of it," as Amelia Earhart said, but Edna Whyte flew to win.

And she did win. Whyte collected close to 130 trophies in all kinds of air races, as well as many achievement awards. She also had one of the longest careers in aviation. During over 60 years of flying, she put in some 34,000 hours of flight time— probably more than any other woman pilot. She was still flying, and teaching others to fly, when she was almost 90 years old. She trained nearly 5,000 pilots.

━━━━━━

In her autobiography, *Rising Above It,* Edna Gardner Whyte wrote, "I must have been competitive from the day I was born." Edna Marvel Gardner was born on November 3, 1902, in Garden City, Minnesota. When she was three, her family moved to Oregon. There her father, Walter Gardner, first tried farming and then worked as a railroad engineer.

Edna's competitiveness and love of speed both showed themselves early. She loved beating neighborhood boys in skating and toboggan races. Roller skating down a hill, she said later, made her feel as if she were flying.

Walter Gardner was killed in a railroad accident when Edna was eight years old, and his death devastated the family. Edna's mother, Myrtle, became ill with tuberculosis and returned to Minnesota. Edna, her brother, and her sister were sent to live with different relatives. "I felt unloved and unwanted . . . as if there was no one I could trust."

Edna's hard childhood developed her determination and self-reliance. She remembered her mother's advice on dealing with difficulty: "Rise above it." When she grew up, Edna would follow that advice in a very literal way.

Edna Gardner graduated from high school in New Salem, Wisconsin, in 1921. Her mother had regained her health by then, and Edna and her brother, Dean, lived with her. Edna then studied nursing. She became a registered nurse in 1924.

Gardner took her first plane ride in 1926, while she and her brother were visiting an aunt and uncle in Seattle, Washington. Bob Martin, a friend she had met there, took her up. Martin was hardly an expert pilot—Gardner later found out he had had only eight hours of flying time—but those rides wedded her to the air for life.

Soon afterward, at her mother's urging, Gardner signed up for premedical courses at the university in Madison, Wisconsin. She also continued her nursing. She did not forget flying, however. How could anyone forget flying in the late 1920s, when the names of Charles Lindbergh and Amelia Earhart were on everyone's lips? Gardner hoped that flying might someday bring her fame as great as theirs.

Gardner saved every spare cent from her nurse's salary to pay for flying lessons. The pilots at the local airport, however, refused to give her much help. They either tried to date her or made fun of her. "Settle down and give some nice guy babies," one told her.

In 1929, partly because of lack of money, Gardner quit her premedical training and joined the U.S. Naval Nurse Corps. She was assigned to the Great Lakes Naval Hospital in Illinois. There she tried again to find someone who would teach her to fly, and this time her luck was better. Her instructor, Guerdon ("Guerd") W. Brocksom, encouraged instead of belittling her.

Gardner first flew solo on New Year's Day, 1931. She had never felt so happy. She later remembered thinking, "Just watch, all of you men. I'll show you what a woman can do. . . . I'll go across the country, I'll race to the moon. . . . I'll never look back."

Gardner took the tests for her private pilot's license in May. She finished the written test more quickly than the two men who were also taking it, and she got a higher score. But at first the sour-faced Department of Commerce inspector refused to take her up for her check ride, which would test her flying performance. "I've never given any woman a [pilot's] license and I'm not at all sure that I want to start now," he said. He asked why Gardner wanted a license.

Gardner's anger and frustration boiled over into tears. "I want . . . a career in aviation. Am I being denied the chance to take the check ride just because I'm a *woman?*"

The inspector gave in and allowed Gardner to take her plane up. He could find no fault with her performance, and he granted her her license—"grudgingly," she says.

Gardner and Guerd Brocksom, her handsome flying instructor, had become more than friends. Soon after she got her pilot's license, however, she had to tell him that the navy had reassigned her to Newport Naval Hospital in Rhode Island. She would have to leave in two weeks. She hoped Brocksom would follow her to Newport.

Gardner bought her first plane, a Travel Air, in Newport. In September 1931, with her private license less than four months old, she flew it to Ohio to take part in the Cleveland Air Races. In the hope of earning back some of the money the trip would cost, she persuaded the Triangle Parachute Company to provide her with a parachute and a jumper so she could enter the parachute jumping contest. The jumpers tried to land inside circles painted on the ground. The most accurate jumper and the jumper's pilot would divide the prize money.

Gardner's jumper was a short, hard-eyed man who called himself Cowboy LaPierre. He landed inside the painted target almost every time. He told Gardner to come to the race office after their last jump to pick up her half of the prize money they were sure to win. When Gardner reached the office, however, LaPierre had already come and gone—with all the money. "That's what you get for thinking you can be a pilot, lady," sneered one bystander.

Brocksom moved to Newport at the beginning of 1932. A few months later, he and Gardner took over the management of a small airfield. They hoped to give sightseeing rides and flying lessons there. Gardner continued nursing as well.

Gardner and Brocksom bought a faster plane and modified it so Gardner could enter air contests as a racer. She won her first air race in Boston in July 1933. After that, she entered every race she could. She told Brocksom that racing was more fun than anything she had ever done.

Whenever Gardner won a race, she made sure the newspapers learned about her triumph. She hoped the publicity would bring more customers to her airfield. Some women pilots, such as Anne Lindbergh, hated publicity, but Gardner, like Bessie Coleman, thrived on it and actively sought it out.

Gardner also earned her instructor's license and began giving flying lessons that summer. She found teaching very rewarding. "Teaching was the best way to learn," she wrote.

The following spring, the navy told Gardner that she had been reassigned to Washington, D.C. Unwilling to make another move, Brocksom asked Gardner to resign from the navy and stay in Newport with him. She said no.

In Washington, Gardner honed her racing skills. She had learned the trick of watching the starter's arm from a fellow pilot in 1931, before she began racing. Now she learned how to calculate her gas exactly: enough to finish the race, but no extra to add weight and slow her down. She learned to reduce weight further by removing all parts of the plane that were not essential for function or safety, such as extra seats. The lighter a plane was, the faster it would go.

Gardner enjoyed and often won cross-country contests, but closed-course pylon racing was her favorite kind of air race. She loved the fierce competitiveness of these brief, intense battles, which reminded her of wrestling. In them she struggled, not only against other pilots, but also against turbulent air and the controls of her own plane as she pushed it into tight turns. The ever present danger of crashes and midair collisions excited rather than frightened her.

Gardner formed "guarded and competitive" friendships with the other women pilots she met on the racing circuit. She enjoyed socializing at the races and at meetings of the Ninety-Nines, the women pilots' association that Amelia Earhart had started. (Gardner was one of the group's charter members and served as its president from 1955 to 1957.) She met Amelia Earhart, Anne Lindbergh, and Jacqueline Cochran. But no friendship ever made her hesitate during a race. "A shrinking violet would not last a day in a career in aviation," she wrote. "If the men didn't browbeat her to death, the other women would."

In Washington, Gardner made a new man friend, Ray Kidd. Unlike the quiet Brocksom, Kidd agreed with Gardner that publicity was important. He had a background in the field and wanted to manage her career. Gardner hoped he could do as well for her as George Putnam had done for Amelia Earhart.

Gardner married Kidd in 1935. She then did what she had refused to do for Brocksom: she resigned from the navy.

The Kidds moved to New Orleans and started a flying school on property they bought at a small airport. After World War II began in Europe in 1939, the U.S. government started the Civilian Pilot Training Program (CPTP) to train pilots who might be needed if America entered the war. The Kidds took advantage of this opportunity, just as Katherine and Marjorie Stinson had done with a similar one before World War I. The Kidds' school, the New Orleans Air College, became part of the CPTP. It began training its first group of pilots on June 15, 1940.

Soon afterward, Ray Kidd suddenly told Edna that their five-year marriage was over. He had fallen in love with a younger woman. "You're busy all the time," he complained. "She *needs* me. You're so capable. . . . So strong."

Edna felt she had to be capable and strong, now more than ever. She kept the airport property and the flying school after her divorce from Kidd, and she concentrated on making the pilot training program a success. When the navy offered to buy the property, however, she accepted.

Edna moved to Fort Worth, Texas, in 1941. She took additional training in flying "blind," using only a plane's instruments for navigation. Then, after the United States entered

World War II, she taught instrument flying to military pilots at a flight school in Oklahoma.

When that training program ended abruptly in 1944, Edna returned to her first career. She spent the last part of the war as a member of the Army Nurse Corps. Stationed in the Philippines, she flew transport planes that carried wounded soldiers to base hospitals.

After the war, Edna returned to Fort Worth and once again worked as a flight instructor. Instructors were in demand because many war veterans, taking advantage of the free education promised by the G.I. Bill, wanted to learn to be pilots. While interviewing other potential instructors, Edna met a tall, attractive pilot-mechanic. Just as she had done long before with Guerd Brocksom, she fell in love. She married George Murphy Whyte in August 1946.

Air racing started up again once the war was over. All through the 1950s and 1960s—which were also her own fifties and sixties—Edna Whyte "raced grooves into the sky" and added to her roomful of trophies. Murphy, as she called her husband, kept her planes in top condition and found even more ways to lighten their weight. The Whytes also continued to teach. In the late 1950s, they opened a flight school and airport business, Aero Enterprises, near Fort Worth.

The Whytes divorced in 1967. Like Ray Kidd, Murphy Whyte had grown tired of having an "absentee wife." Edna, for her part, was glad to regain her independence. Murphy did not want to leave Aero Enterprises, but she wanted to start an airport of her own.

In 1969, Edna Whyte (who kept her second husband's name) bought land in Roanoke, Texas, about 18 miles from Fort Worth. The area had been rural, but it was growing rapidly. Whyte was sure an airport in that location would be a success. In 1972, however, when she applied for a loan to pay for building one, the Small Business Administration turned her down. At 70 years of age, the agency said, she was too old for such a venture. Furthermore, managing an airport was supposed to be a man's job.

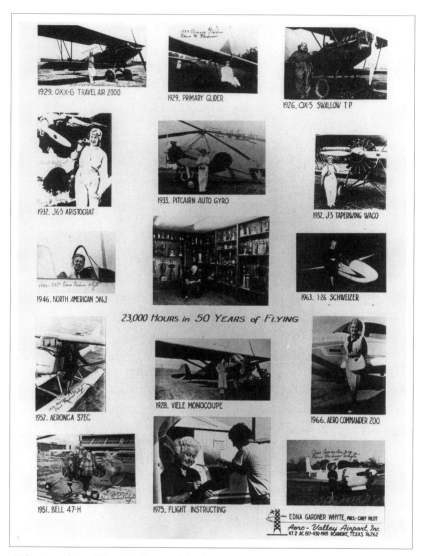

Edna Gardner Whyte had one of the longest careers in aviation. This picture shows some of the planes she flew during the first 50 years of that career. By the time of her death, Whyte had flown for over 60 years and had put in about 34,000 hours of flight time.
(Courtesy National Air and Space Museum, Smithsonian Institution, SI Neg No. 76-2186)

As usual when presented with such obstacles, Whyte simply became more determined. With the help of Kelly Bryan, a young man who became a close friend, she designed Aero Valley Airport and built it with money she raised herself. "Friendly Aero Valley Airport—Growing Without Federal Money!" Whyte's business cards proclaimed.

During the 1970s, Whyte lived in an apartment built into the rear of one of the airport's hangars. In addition to managing the airport, she opened another flight school. In noisy trainer planes, she captured her students' attention by pulling on their earlobes. She liked to tell the story of a doctor in Fort Worth who supposedly said to a patient, "Oh, you must be one of Edna Whyte's flying students." The surprised man asked how the doctor had guessed. "One of your earlobes is longer than the other," the doctor replied.

Whyte remained adventurous to the end of her life. She had always wanted to fly a military jet, and in 1983 she got her chance. After a young air force captain took her up in a jet trainer, he told the 81-year-old woman, "I could make a fighter pilot out of you." Whyte said it was the greatest compliment she had ever received.

Whyte sold most of her stake in Aero Valley Airport in 1980, but she continued to live and work there. By 1985, the airport had hangar space for 360 planes, housed three flight schools, and specialized in restoring antique planes. The airport's success allowed Whyte to replace her hangar apartment with a full-sized house. Instead of a garage, it had a hangar for her Cessna-120. A paved taxiway led from Whyte's front door to the airport's main runway.

Whyte's age did not slow her down. Even in her eighties, she entered races several times a year and often won. She got up at dawn, taught both day and night flying, gave license examinations for the Federal Aeronautics Administration, and remained strong enough to open and close heavy hangar doors with ease. Whyte wrote near the end of her autobiography, "I want to fly until I'm 100," and she came close to achieving that goal. She died on February 15, 1993, at age 91.

In addition to her racing trophies, Edna Gardner Whyte won many awards during her long career. She was elected to four halls of fame, including the Aviation Hall of Fame, and won the Charles Lindbergh Lifetime Achievement Award. The Pioneer Pilots' Association gave her an award for being "the woman who contributed most to aviation."

In her later years, Whyte rejoiced in the achievements of younger women such as Bonnie Tiburzi and Sally Ride. She never got over her regret that she had not had a chance to make similar achievements, but she was proud of her work as a racer and a teacher. She wrote, "I hope that I have helped hold open the door to [the] sky so that every woman can 'rise above it.'"

Chronology

November 3, 1902	Edna Marvel Gardner born in Garden City, Minnesota
1916	graduates from New Salem High School in Wisconsin
1924	becomes a registered nurse
1926	takes first plane ride
1929	joins U.S. Naval Nurse Corps
1930	takes flying lessons from Guerd Brocksom
May 1931	earns private pilot's license
September 1931	enters first air contest
July 1933	wins first air race; begins giving flying lessons
1935	marries Ray Kidd
June 15, 1940	flying school begins training pilots as part of Civilian Pilot Training Program 1940divorced from Ray Kidd
1941	takes instrument training and begins teaching in war program
1944	joins Army Nurse Corps
August 1946	marries George Murphy Whyte
late 1950s	Whytes open Aero Enterprises
1967	divorced from George Murphy Whyte
1969	buys land for airport in Roanoke, Texas

1972	denied government loan because of age; builds Aero Valley Airport anyway
1983	flies in jet trainer
February 15, 1993	Edna Gardner Whyte dies

Further Reading

Baxter, Gordon. "Iron Edna." *Flying,* May 1984. Brief review of Whyte's flying accomplishments.

Blount, Joan Ackerman. "She Flies Through the Air with the Greatest of Ease," *Sports Illustrated,* January 16, 1984. Long article includes an interview with Whyte and a review of her career.

O'Connor, Diane Chamness. "Flying High." *Good Housekeeping,* September 1985. Reviews Whyte's life and describes her activities in her eighties.

Whyte, Edna Gardner, with Ann L. Cooper. *Rising Above It.* New York: Orion Books, 1991. Whyte's autobiography.

Anne Morrow Lindbergh
(1906–)

Anne Morrow Lindbergh was a pilot in her own right as well as an able copilot for her famous husband, Charles Lindbergh. Her writings help readers experience the beauty and terror of flying.
(Courtesy National Air and Space Museum, Smithsonian Institution, SI Neg No. A45256-E)

*A*s the Lockheed Sirius edged past the mountain on the small island between Russia and Japan, fingers of fog curled around the plane. Her husband struggled to control the craft while Anne Lindbergh watched the growing mist bring blindness— and perhaps death:

> *I was losing the sky. I did not want to let go until I could grasp something below. Down the sides of the mountain one could see a strip of water gleaming. . . . Hold onto the sea—that little patch*

of blue. Oh, the sea was gone too. We were blind—and still going down— oh, God!—we'll hit the mountain! . . . Then a lurch . . . and a sickening roller coaster up. . . . Thank God, there is the sky. Hold on to it with both hands. Let it pull you up. Oh, let us stay here, I thought, up in this clear bright world of reality, where we can see the sky and feel the sun. Let's never go down.

Anne Morrow Lindbergh and her famous husband, Charles, encountered many heart-stopping moments during their flight over the Arctic to Asia in 1931. Anne Lindbergh described them in *North to the Orient,* her book about the trip. She became a competent pilot and skilled radio operator, but her greatest gift to aviation was her power to capture in words the beauty and terror of being in the air.

Anne Spencer Morrow was born on June 22, 1906, in Englewood, New Jersey. She and her two sisters, Elisabeth and Constance, and brother, Dwight Jr., led a quiet, protected existence as the children of wealthy parents. Dwight Morrow, their father, was a partner in the powerful financial firm of J. P. Morgan & Co. He wanted his children to have all the privileges that he and his wife, Betty, had lacked in their own modest childhoods in the Midwest. The Morrows toured Europe often, and the children attended expensive private schools. Anne's early life, she once told an interviewer, "consisted of family gatherings, walks in the woods, books and thinking."

After Anne graduated from high school, her parents sent her to Smith College in Northampton, Massachusetts. Her mother and older sister also had gone to college there. Anne was small and pretty, but she was very shy and seldom dated during college. She played basketball and, according to biographer Joyce Milton, was "a far better athlete than she gave herself credit for." Her main interest, however, was creative writing. She won Smith's highest literary award, the Mary Augusta Jordan prize, and one of her poems was published in a national magazine before she graduated from Smith in 1928. Interestingly, in that poem she imagined flying as an escape from earthly cares.

In high school, Anne Morrow had written in her yearbook, "I want to marry a hero." But when she had a chance to meet a hero, in December 1927, she felt only irritation. Her father had recently been made U.S. ambassador to Mexico, and he invited Charles Lindbergh, "Lucky Lindy"—known everywhere after his solo crossing of the Atlantic earlier that year—to spend Christmas in Mexico City. Lindbergh was to stay with the Morrows during the visit, and Anne was sure he would ruin her Christmas vacation. She expected him to be a "baseball player" type who knew nothing about books.

Once she met Lindbergh, Morrow's feelings changed. She wrote in her diary, "This clear, direct, straight boy swept out of sight all the other men I have ever known." Lindbergh was also impressed with the quiet Anne. "He told me years later that he was attracted to me because I was the only girl who hadn't flirted with him," she recalled. In the summer of 1928, when he was again in Mexico, Lindbergh asked Morrow to go up with him for a flying lesson. It was his first date.

Morrow and Lindbergh were alike in being shy, private people. In other ways, however, they were very different. Anne was an observer; Charles was a doer. "You want to *write!*" he once exclaimed. "I want to do the things other people write about!" Morrow hoped he could teach her to be more active. "He filled some kind of hunger in me to break out of the pattern which was neatly laid out before me," she later told an interviewer. "I suddenly felt that I had been living in dreams and that he was life itself."

When Morrow's parents announced her engagement to Lindbergh early in 1929, newspapers called it "the romance of the century." Reporters and photographers followed them everywhere. To avoid them, the couple's wedding at the Morrows' New Jersey mansion on May 27, 1929, was kept secret. The Lindberghs spent their honeymoon on a motorboat, cruising the Maine coast, but even there reporters pursued them.

Charles Lindbergh was hailed as a hero after he became the first person to fly alone across the Atlantic in 1927. When Anne Morrow met him, she expected to dislike him; instead, she fell in love with him.
(Courtesy National Air and Space Museum, Smithsonian Institution, SI Neg No. A4812-B)

Similar press attention dogged the Lindberghs most of their lives. The harder they tried to hide, the more they were hunted down. If they wanted to go out, they had to put on disguises such as false eyeglasses and heavy makeup. Anne said she felt like an escaped criminal.

Anne let Charles teach her to fly so she could share his adventurous life. His demands for perfect performance sometimes left her "challenged, frightened, and infuriated," but she came to enjoy being in the air. She once described flying as "beauty, adventure, discovery." She earned her pilot's license early in 1931. Before that, in 1930, she learned to fly a glider, or sailplane, and became the first American woman to get a glider pilot's license. She also flew with her husband as copilot and navigator when he set a speed record between Los Angeles and New York in April 1930. Two months later, on her twenty-fourth birthday, Anne gave birth to her first child, Charles Augustus Lindbergh, Jr.

Charles Lindbergh wanted to promote commercial aviation, both in the United States and overseas. "The thing that interests me," he told Anne, "is breaking up the prejudices between nations, linking them up through aviation." He worked as a consultant for Pan American Airways and other airlines. Once Anne was able to fly with him, he began planning several long survey flights. These flights would help airlines determine the best routes for future travel.

The first route the Lindberghs explored was the "Great Circle" route from the United States to Asia. Ships already used this approach, which went north past Canada and Alaska rather than west across the Pacific. If planes could use it as well, it would save time, fuel, and money. The Lindberghs flew in a Lockheed Sirius, an open-cockpit monoplane (plane with one set of wings). It had been given an extra-powerful engine, additional gas tanks, and pontoons so it could land in the water. It carried an assortment of survival gear, ranging from a rubber boat to an insect-proof tent. To prepare for the trip, Anne learned navigation and radio operation. She wrote that radio work at first seemed to require four hands: one to control the main dial, one to tune the station, one to write the message, and one to steady the writing pad.

Charles and Anne Lindbergh flew over the Arctic to Asia in their Lockheed
Sirius in 1931. Anne Lindbergh later described the flight in a book,
North to the Orient.
(Courtesy National Air and Space Museum, Smithsonian Institution,
SI Neg No. 13 A4818-B)

Leaving their year-old son in the care of Betty Morrow and
a nursemaid, the Lindberghs began their Arctic flight from
New York on July 27, 1931. That evening, they reached Ottawa,
Canada. There experienced pilots tried to persuade them not

to fly further along the desolate route. "I wouldn't take *my* wife into that territory," one pilot said.

"You must remember," Charles Lindbergh answered, "that she is *crew*." Anne Lindbergh wrote later that this moment, in which her husband accepted her as a flying equal, was the proudest of her life.

After three forced landings and the narrow escape from the mountainous island, the Lindberghs reached Japan. They then went on to China, where the Yangtze River was having one of its periodic floods. They saw thousands of homeless villagers crowded onto small boats called sampans, trying to survive until the waters ebbed.

In Nanking, a famine relief team asked the Lindberghs to make an air survey of the flood. No other plane in China had enough range to cover the whole area. During the survey, Anne flew the plane while Charles drew sketches and maps. They also helped a Chinese doctor deliver medical supplies. When they landed, the starving peasants, believing they had brought food, mobbed them and almost sank the plane.

After the Lindberghs came back to the United States, their baby son was stolen from their home on the night of March 1, 1932. The kidnapping became front-page news, and over 28,000 people wrote to offer sympathy and suggestions.

The baby's body was found in the woods near the Lindbergh house on May 12. Charles Lindbergh became obsessed with finding his son's killers, but Anne, who was pregnant again, wanted to forget the unhappy past. Her second son, Jon, was born on August 12.

In late July 1933, the Lindberghs began a long-distance flight to explore northern and southern routes across the Atlantic to Europe. Again they flew their Lockheed Sirius, which they had now named *Tingmissartoq*. This Eskimo word means "flies like a big bird."

The trip began with an air crossing of the Greenland ice cap, which had been done only twice before. Sometimes the Lindberghs flew through snowstorms and fog. When the air was clear, they faced a wasteland of snow whose glitter could cause

Anne Morrow Lindbergh

temporary blindness. After Greenland, they went on to Iceland, the Faroe Islands, and Denmark.

The Lindberghs toured Europe, then tried to fly back across the Atlantic by the southern route. They found that harbors and weather in both the Azores and the Canary Islands, which they had hoped to use as stops, were unsuitable for seaplanes. They finally left from Bathurst, Gambia (Africa), after a struggle that involved dumping 200 pounds of supplies and waiting for three days for the wind to rise and help them on their way. In *Listen! The Wind,* her book about the trip, Anne described the moment of takeoff:

> *It [the sound of the plane's engine] smooths out now, like a long sigh, like a person breathing easily, freely. Like someone singing ecstatically, climbing, soaring—sustained note of power and joy. . . . We did it, we did it! We're up, above you. We were dependent on you just now, River, prisoners fawning on you for favors, for wind and light. But now, we are free.*

After stops in South America, Central America, and the Caribbean, the Lindberghs returned to New York on December 19. Charles said that the five-and-a-half-month journey was the high point of his flying career. Anne, too, won awards for the trip, including the National Geographic Society's Hubbard Medal. She was the first woman to receive this medal. (The society had earlier given Amelia Earhart a different award.) The medal, given in 1934, honored Anne for "courageous and skillful work as copilot and radio operator during 40,000 miles of flight over five continents and many seas."

The Atlantic trip had taken the Lindberghs away from the memory of their baby's death. The memory returned on September 20, 1934, when Bruno Richard Hauptmann, a German-born carpenter, was arrested and charged with the kidnapping. Charles was delighted with the news, but Anne wailed, "It's starting all over again!" Charles went to the courtroom every day during Hauptmann's trial. Anne,

57

however, went only twice, when she had to testify. Hauptmann was found guilty and later executed.

Anne turned her notes about the Great Circle trip into a book, *North to the Orient*. When a publisher accepted it, she feared at first that this had happened only because of her famous name. Her editor, however, assured her, "I would have accepted the manuscript if it had been written by Jane Smith." The book was published in 1935 and sold well.

Still fleeing the publicity that continued to haunt them, the Lindberghs moved to England in December 1935. There Anne gave birth to a third son, Land, in 1937. She also wrote a second book, which described the last part of the Atlantic flight. It was published in 1938 as *Listen! The Wind*.

In the summer of 1936, the Lindberghs accepted an invitation from Hermann Goering, the head of the German air force, to inspect that country's air installations. The American military attaché in Berlin asked Charles Lindbergh to make the visit. The attaché hoped that Lindbergh might learn something useful to the American government, which was concerned about the Nazis' military buildup.

Following this trip and two others in 1937 and 1938, Lindbergh concluded that Germany's air strength was greater than it really was. He also was impressed with the Nazi government's apparent efficiency. He said that Britain and France should let Hitler take over the territory he wanted in eastern Europe. He believed they had no hope of defeating Germany in a war.

At a dinner in Berlin in October 1938, Lindbergh accepted the Order of the German Eagle from Goering. For a while, the Lindberghs even considered moving to Germany. They dropped all such plans, however, when they realized the extent of the Nazi government's persecution of the Jews.

Realizing that war was likely to begin in Europe at any moment, the Lindberghs returned to the United States in May 1939. They received a cold reception because of Charles Lindbergh's apparently pro-Nazi views. Lindbergh said he hoped the United States would stay out of what he saw as a strictly European quarrel, and Anne supported him. Once the United States entered the war, however, Lindbergh immediately tried to enlist. He was never officially accepted into the military, but he flew 50 bombing missions in the Pacific.

During the late 1940s and 1950s, Charles Lindbergh continued to work as a consultant in military and civil aviation. He traveled often, but Anne stayed home in Darien, Connecticut, to raise their large family. That family now included two daughters (Anne, born in 1939, and Reeve, born in 1946) and a fourth son (Scott, born in 1942). Anne wrote several more books, including two novels and a book of poetry. None was about flying. Only one of these later books, *Gift from the Sea* (1955), sold well. It was a group of essays on the meaning of being a woman and the need for quiet times alone during even the busiest life. It is still in print and has sold over seven million copies.

In the 1960s, the Lindberghs became concerned about the destruction of Earth's environment. They traveled to Africa and the Philippines and worked with the World Wildlife Fund and other groups to protect wildlife and native human cultures. In one speech, Anne said that "human values spring from earth values and must be supported by them." She wrote a magazine article describing a trip to Africa, during which she unexpectedly encountered two lions. "I feel honored that they have not considered me worth running away from—or towards," she wrote.

Both Lindberghs published selections from their extensive diaries in the late 1960s and early 1970s. Anne's five collections of diaries and letters, which covered her life from her youth to 1944, were especially well received. Charles Lindbergh died of cancer in 1974. Anne continued to write until the end of the 1970s, when her health failed.

Anne Lindbergh's main importance as a woman aviator is her ability to capture in words the "fundamental magic of flying." As she wrote in *North to the Orient*, "Flying, like a glass-bottomed bucket, can give you that vision, that seeing eye, which peers down to the still world below the choppy waves." She referred not just to physical water but to the fundamental values that lie beneath the confusions of life. Lindbergh always looked for these values. One newspaper article described her as "a rare combination of daring adventuress and introspective seeker of truth."

Chronology

June 22, 1906	Anne Spencer Morrow born in Englewood, New Jersey
December 1927	meets Charles Lindbergh
1928	graduates from Smith College
May 27, 1929	marries Charles Lindbergh
1930	becomes first woman to earn glider pilot's license; accompanies husband on record-setting transcontinental flight
1931	earns private pilot's license
July 27, 1931	begins flight across Arctic to Asia
March 1, 1932	baby son kidnapped
July 1933	begins flight to survey transatlantic air routes
1934	wins Hubbard medal
1935	*North to the Orient* published; Lindberghs move to England
1936	Lindberghs pay first visit to Germany
1938	*Listen! The Wind* published; Charles Lindbergh accepts German medal
May 1939	Lindberghs return to United States
1955	*Gift from the Sea* published
1960s	Lindberghs work for protection of environment; publish diaries
1974	Charles Lindbergh dies

Further Reading

Boase, Wendy. *The Sky's the Limit*. New York: Macmillan, 1979. Includes a chapter on Lindbergh.

Genett, Ann. *Contributions of Women: Aviation*. Minneapolis, Minn.: Dillon Press, 1975. For young adults. Chapter on Lindbergh includes some long quotes from her works.

Herrmann, Dorothy. *A Gift for Life*. Boston: Ticknor & Fields/Houghton Mifflin, 1992. Biography of Anne Morrow Lindbergh.

Lindbergh, Anne Morrow. "A Brief Safari Back to Innocence." *Life*, October 21, 1966. Describes visit to Africa and impressions of African wildlife.

———. *Listen! The Wind*. New York: Harcourt, Brace, 1938. Describes the last part of the Lindberghs' transatlantic trip.

———. *North to the Orient*. New York: Harcourt, Brace, 1935. Describes the Lindberghs' Great Circle flight over the Arctic to Asia.

"Lindbergh, Anne (Spencer) Morrow." *Current Biography*, 1976. Concise but interesting sketch of Lindbergh's life and career.

Lomax, Judy. *Women of the Air*. New York: Ivy/Ballantine, 1988. Includes a chapter on Lindbergh.

Milton, Joyce. *Loss of Eden*. New York: HarperCollins, 1993. Biography of Charles and Anne Lindbergh.

Thrush, Robin A. "A Hero's Wife Remembers." *Good Housekeeping*, June 1977. Interesting interview in which Anne Lindbergh recalls highlights of her marriage and career.

Jacqueline Cochran
(1906?–1980)

Jacqueline Cochran broke many barriers during her career as an air racer and test pilot. In 1953, she became the first woman to fly faster than the speed of sound.
(Courtesy National Air and Space Museum, Smithsonian Institution, SI Neg No. 72-6626)

*U*p, up, up climbed the F-86 Sabrejet. At last it reached 45,000 feet. At that height, even though daylight covered the earth below, the test pilot in the Canadian military jet could see stars shining brilliantly.

In a gentle curve at first, then more steeply, the pilot pointed the jet's nose downward. Soon the plane was streaking straight toward the ground—at 500, 600 miles an hour, and more.

"Ninety-seven . . . 98 . . . 99 . . . " the pilot radioed to the chase plane close behind. Then, suddenly, the noise of the jet died.

The test pilot knew the jet's engines had not quit. Instead, a barrier had been broken. The jet had exceeded Mach 1, the speed of sound. The sound waves carrying the noise of the plane could not keep up with the plane itself.

Jacqueline Cochran said that this moment in May 1953, when she became the first woman to break the sound barrier, was the greatest thrill of her life. But the speed of sound was only one of many barriers Cochran broke during her long career in the air. As Floyd Odlum, Cochan's husband, wrote of her, "Certain it is that she is fearless of death and equally certain it is that she considers a barrier only something to surmount."

Jackie Cochran broke her first barrier while she was little more than a child: the barrier of her background. She guessed she had been born near Muscogee, Florida, around 1906, and she celebrated her birthday on May 11, but she never knew for certain when or where she had been born or who her parents were. She chose her last name, *Cochran,* from a telephone book when she was a teenager.

Like Bessie Coleman and Edna Gardner Whyte, Jackie Cochran grew up in poverty. She was raised by a foster family of sawmill workers in northern Florida. They traveled from mill camp to mill camp along what Cochran later called "Sawdust Road." Jackie slept on the floor and wore dresses made from discarded flour sacks. Cochran wrote later that her deprived childhood gave her "a kind of cocky confidence. . . . *I could never have so little that I hadn't had less. It took away my fear."

Young Jackie was delighted when she heard her foster mother tell a neighbor that Jackie was not her real daughter. "I was glad I wasn't related by blood to those shiftless people," Cochran later said. "Just knowing I wasn't really one of them gave me incentive to . . . improve my lot."

Jacqueline Cochran

Hoping to earn enough money to escape Sawdust Road, Jackie took any job she could find. Before she was 10, she was supervising 15 other child workers in a cotton mill in Columbus, Georgia. She told her unbelieving workmates that she would someday be rich: "I'll wear fine clothes, own my own automobile, and have adventures all over the world."

When a strike ended her mill job, Jackie persuaded the owner of a beauty shop in Columbus to hire her as an assistant. She worked as a beauty operator in larger and larger cities until, in 1929, she reached New York City. There she got a job at Antoine's, a popular beauty salon that had many wealthy customers.

Some of Cochran's clients became her friends and invited her to their parties. At one such party in 1932, she saw a man who attracted her immediately, even though he was older than she (14 years older, she later learned). When the two were introduced, the attraction proved to be mutual. Only later did Cochran discover that Floyd Odlum was a multimillionaire businessman.

Odlum and Cochran became friends. She told him she wanted to be a traveling cosmetics saleswoman. Odlum said that to make a living that way during the depression, she would have to travel so much that she would need wings.

Cochran took Odlum's phrase literally. In the summer of 1932, she decided to spend part of an upcoming six-week vacation at the Roosevelt Flying School on Long Island. Odlum bet her the cost of her lessons that she could not qualify for her pilot's license within the six weeks. Cochran said she would do it in three, because she did not want to spend her whole vacation at the flight school. To Odlum's amazement, she succeeded. She then took further lessons at another school. By the end of 1933, she had earned her commercial pilot's license and instructor's rating.

Jackie Cochran often said that when she took her first flying lesson, "A beauty operator ceased to exist and an aviator was born." Only the second part of the statement was completely true. Cochran still wanted to sell beauty products—but not for someone else. In 1935, with Floyd Odlum's help, she started the Jacqueline Cochran Cosmetics Company. She designed many of its products herself, including hair dyes, moisturizing creams, and cream sticks to soothe chapped lips. The company was so successful that a poll of newspaper editors named Cochran "Businesswoman of the Year" in 1953 and 1954. Cochran sold her interest in the company in 1964, but the firm still existed in the 1980s.

Cochran continued to grow closer to Floyd Odlum, and they married on May 11, 1936. Although Cochran, like Amelia Earhart, continued to use her maiden name professionally, her marriage to Odlum was a lifelong and happy match. Odlum admired Cochran's energy and encouraged her in everything she did. Severe arthritis disabled him while he was still in his forties, and after that his more adventurous side expressed itself mainly through Cochran's achievements. Cochran, in turn, was devoted to him. She depended on him for money, contacts with people in high places, and sensible advice. One friend said, "Theirs is one of the great love affairs of all time. Everything she does revolves around him, and he is enthralled by whatever she is doing." The two bought land in Indio, in the Southern California desert, and spent most of their time in a large ranch house there. Cochran helped build the house and decorated it herself.

Cochran began entering air races and setting aviation re-cords in 1934. Within a few years, she had joined Amelia Earhart, Anne Lindbergh, and Edna Whyte as one of the top women pilots in the world. In 1937, the year of Earhart's death, for example, Cochran set three major records: a speed record between New York and Miami, a women's national speed record, and a women's world speed record. In 1938, Cochran won the Harmon Trophy, awarded by the International League of Aviators to the year's best woman pilot.

Although Cochran, Earhart, and Whyte all knew each other—Cochran and Earhart, indeed, were close friends—they seldom competed directly. Each had a different specialty. Ear-hart was a long-distance flier. Whyte was a racer. Cochran, by contrast, saw herself primarily as a test pilot. "The objective in every one of my flights," she wrote, "was to go faster or farther through the atmosphere or higher into it than anyone else and to bring back some new information about plane, engine, fuel, instruments, air, or pilot that would be helpful in the conquest of the atmosphere."

Because Cochran saw her flying as a form of aviation research, she supported others in the same field. In 1937, she persuaded a committee to award the Collier Trophy, which was supposed to go to someone who had made a major accomplishment in aviation, to a young Mayo Clinic surgeon named Randolph Lovelace. Working in their spare time, Lovelace and an assistant had developed an oxygen mask and tank for pilots flying at high altitudes.

Cochran knew from her own high-altitude flights that low air pressure and a shortage of oxygen above 20,000 feet could make a pilot bleed from the nose, have hallucinations, or even pass out. Cockpits should be pressurized, she believed, or else pilots should breathe extra oxygen from tanks and masks like the one Lovelace had invented. Giving the Collier trophy to Lovelace would help him get the recognition and funding he needed to continue his work, Cochran told the committee. That work would be especially important if, as many people predicted, war with Germany broke out and eventually involved the United States. Protecting combat pilots at high altitudes would then be a major concern.

Although Cochran, like Edna Whyte, was very competitive, Cochran entered air races mainly as a way of testing planes. Her favorite challenge was the Bendix cross-country air race, which was held yearly around Labor Day. The racers flew between Los Angeles and Cleveland, Ohio. In its time, Cochran has said, the Bendix was the outstanding regular long-distance air race of the world.

Cochran succeeded in winning the Bendix in 1938. She flew a Seversky P-35 pursuit plane, a new military plane that was considered fast but unreliable.

Cochran's P-35 had gas tanks in its wings. Fuel was supposed to feed from both tanks equally to keep the plane balanced. As

Cochran flew, however, she found that the plane's right side was becoming heavier than its left. Gas, she guessed, was feeding from only the left tank.

Finally the left tank ran dry. The P-35's engine stopped for lack of fuel, and the plane began to spiral out of the sky. Cochran struggled to regain control. She then flew with her left wing down for a while so that gravity would drain gas from the right wing tank into the center tank. She had to repeat this maneuver from time to time throughout the race. She later found that a piece of paper had blocked the outlet of the right wing tank.

Cochran won the Bendix in spite of her gas tank problems, but even that was not enough for her. As soon as she had received her trophy and refueled her plane, she set off for New York City. By continuing her flight, she set a new women's record for west-to-east transcontinental flight in a propeller-driven plane: 10 hours, 27 minutes, 55 seconds.

War erupted between Germany and the Allies, including Britain and France, in 1939. The U.S. government sent help, including warplanes, to Britain, but this had to be done through Canada because the United States was officially a neutral country. On June 17, 1941, Cochran flew a Hudson V bomber from Canada to Britain to publicize Britain's need for pilots and planes. This flight made Cochran the first woman to fly a warplane across the Atlantic. (She did not fly it completely alone, however; the government insisted that a male copilot perform the takeoff and landing.)

In Cochran's opinion, her bomber flight did something far more important than setting another record. It showed that women pilots could handle noncombat military flying. If women were allowed to do such work, she said, male pilots could be freed for combat duty in wartime. She learned that women pilots in Britain were already ferrying military planes from one airport to another as part of the British Air Transport Auxiliary (ATA). She urged the U.S. government to set up a similar arrangement.

The ATA asked Cochran to look for American women pilots who would volunteer to join its program. After careful screening, she recruited 25 "ATA-girls" and took them to England. Some of the British, living under war conditions that bordered on poverty, resented the flamboyant Cochran appearing at airports in an expensive car and wearing a mink coat. No one could deny the excellent performance of her young women, however. Although they signed up to serve for only 18 months, several stayed for the entire war.

While Cochran was in Britain with the ATA-girls, the U.S. military took her advice and set up an organization of women pilots to ferry planes and carry out other noncombat flying duties. There was only one problem, as far as Cochran was concerned: they put someone else in charge of it. She respected Nancy Love, the woman pilot the government chose, but she felt that she herself should have been given the job.

General H. H. "Hap" Arnold, a friend of Cochran's, offered her the chance to head a similar group. Love's small group, the Women's Auxiliary Flying Service (WAFS), were already experienced pilots. Cochran's group would be much larger and would include women with relatively little flying experience. An important part of Cochran's job would be setting up a training program for them.

Cochran eventually accepted 1,830 women out of 25,000 applicants. Of those accepted, 1,074 successfully completed the difficult 23-week training. This rate of success was about the same as that for male Army Air Force cadets. In the spring of 1943, Cochran told her first class of trainees that they were being given "the greatest opportunity ever offered women pilots anywhere in the world."

Cochran's and Love's groups were merged on August 5, 1943, to form the Women's Airforce Service Pilots (WASPs). Cochran was named director of women pilots. Love still managed her own group, but she was subordinate to Cochran.

WASPs learned to fly almost every plane used by the Army Air Forces, including huge B-29 Superfortresses and speedy Mustang and Thunderbolt fighters. One WASP, Ann Carl,

Jacqueline Cochran was director of women pilots for the Women's Airforce Service Pilots (WASPs) during World War II. Here Cochran pins wings on a student who has just completed the challenging WASP training program. (U.S. Air Force Photo Collection [USAF Neg No. A-36492 AC], courtesy of National Air and Space Museum, Smithsonian Institution)

became the first American woman to test-fly a jet fighter. Some WASPs ferried as many as 10 planes a day. Others towed targets for anti-aircraft gunnery practice. The target was a long cloth sleeve tied by a long rope to the tail of the plane the WASP flew. Still other WASPs made low-altitude flights to give radar and searchlight operators practice in spotting them. Together the WASPs flew some 60,000 hours and 60 million miles. They delivered 12,650 planes of 77 different types. They performed as competently as male pilots, with less time lost to illness and accidents.

To Cochran's (and most WASPs') fury and regret, Congress cancelled the WASP program on December 20, 1944. Worse still, the group had never been made an official part of the military, so WASPs were not eligible for veterans' benefits. Only on September 20, 1977, did Congress reverse itself and allow the WASPs or their families to receive benefits.

Cochran received her own recognition much sooner. For her work as founder and director of the WASP program, she was awarded the Distinguished Service Medal in 1945. This is the second-highest honor that a U.S. civilian can receive.

After her war work was over, Cochran went back to testing new, fast planes. She became a friend of Charles (Chuck) Yeager, the famous test pilot who in 1947 became the first person to fly faster than the speed of sound. (The speed of sound varies with altitude and air temperature but is about 700 miles per hour.) She wanted to test military jets, just as Yeager was doing. She was told, however, that only air force members on active status, a group that included no women, were allowed to fly air force planes.

The persistent Cochran soon found a way around that barrier. Floyd Odlum owned the company that made the air force's F-86 Sabrejets. He knew that the Canadian Air Force also owned some of these planes. He persuaded Air Canada, which sold the planes, to make Cochran a company pilot. Air Canada could then ask the Canadian Air Force to let Cochran borrow one of its Sabrejets, take the plane back to the United States, and test-fly it.

Chuck Yeager taught Cochran how to fly the high-powered Sabrejet. He followed her in a chase plane when she used the jet to break the sound barrier in May 1953. She made 13 flights in the plane, totaling six hours of air time. During that time, she broke three men's records and flew faster than the speed of sound three times. (In 1964, she flew an even faster jet, the Lockheed F-104 Starfighter, at twice the speed of sound.) She said breaking the sound barrier was like "flying inside an explosion." But, as Jerrie Cobb would do later, Cochran also said that flying very fast and very high was an almost religious experience, creating a feeling of "humility and trust."

Cochran's desire to go higher and faster than anyone else even reached into space. Dr. Randolph Lovelace, the surgeon for whom she had secured the Collier Trophy, worked for the National Aeronautics and Space Administration (NASA) during the early days of the U.S. manned space program. In 1960, Lovelace began a private program to find out whether experienced women pilots could pass the physical and mental tests given to the male astronauts. Cochran paid part of the expenses of Jerrie Cobb and the other 12 women who took the tests. She even took some of the tests herself. "I'd have given my right eye to be an astronaut," she wrote.

Ill health put an end to Jackie Cochran's career in 1971. She had fainting spells that proved to be a form of heart attack. Doctors installed a pacemaker in her chest to steady her heartbeat—and told her she would have to give up flying. When Cochran heard that, a friend said, "Her whole roof caved in." "To live without risk for me would be tantamount to death," Cochran once said. Perhaps because she was denied the risks and thrills of the flying she loved so much, her health quickly declined further.

Other tragedies dimmed Cochran's last years as well. Floyd Odlum's health also grew worse. Arthritis now confined him to his bed. In addition, he made several bad investments, and his losses forced him and Cochran to sell their beloved ranch. Odlum died in 1976 at the age of 84. Cochran outlived him by only a few years, dying in Indio on August 9, 1980.

Jackie Cochran set records as dependably as Edna Whyte won races. At her death, Cochran held some 250 speed, altitude, and distance records—more than any other pilot, male or female, in history. She won many awards as well, including the Distinguished Flying Cross (1969) and the Order of Merit (1970) from the U.S. government. She won the Harmon Trophy for best woman pilot of the year 15 times and was given a special trophy by the Harmon committee—"Pilot of the Decade" 1940–49. She was president of the Ninety-Nines from

1940 to 1943 and of the Fédération Aéronautique Internationale from 1958 to 1961. In 1953, the year she broke the sound barrier, the FAI awarded her its Gold Medal—the first time it gave this award to a woman. Cochran was given honorary wings by the air forces of half a dozen countries, and several colleges awarded her honorary degrees. In 1971, she was enshrined in the Aviation Hall of Fame, the first living woman to receive this honor.

In a 1954 article for *Guideposts* magazine, Jackie Cochran wrote, "I have found adventure in flying, in world travel, in business, and even close at hand . . . Adventure is a state of mind—and spirit. It comes with faith, for with complete faith there is no fear of what faces you in life or death." Cochran's spirit of adventure led her to constantly "push the envelope," as test pilots say. She extended the limits and broke the barriers, not only of aircraft performance, but of her own life.

Chronology

May 11, 1906?	Jacqueline Cochran born in Florida
1932	meets Floyd Odlum; earns private pilot's license in three weeks
1935	starts Jacqueline Cochran Cosmetics Corporation
May 11, 1936	marries Floyd Odlum
1937	sets three major air records; persuades committee to give Collier Trophy to Dr. Randolph Lovelace
1938	wins first Harmon Trophy for woman pilot of the year; wins Bendix air race; sets women's transcontinental speed record
June 17, 1941	becomes first woman to fly a warplane across the Atlantic; recruits ATA-girls
1943	sets up Women's Airforce Service Pilots (WASPs)
December 20, 1944	Congress discontinues WASP program
1945	Cochran awarded Distinguished Service Medal
May 1953	becomes first woman to fly faster than speed of sound; wins FAI Gold Medal
1960	supports testing of women pilots as possible astronauts
1964	sells interest in cosmetics company; flies twice as fast as speed of sound
1969	awarded Distinguished Flying Cross

1971	gives up flying because of ill health; enshrined in Aviation Hall of Fame
1976	Floyd Odlum dies
August 9, 1980	Jacqueline Cochran dies

Further Reading

Brinley, Maryann Bucknum. *Jackie Cochran*. New York: Bantam, 1987. Cochran's autobiography (incorporating an earlier autobiography, *The Stars at Noon*). Includes additional interviews with people who knew Cochran.

"Cochran, Jacqueline." *Current Biography 1963.* Describes Cochran's life and career to this point.

Douglas, Deborah G. *United States Women in Aviation 1940–1985.* Washington, D.C.: Smithsonian Institution Press, 1990. Includes extensive material on Cochran and the WASPs.

Hodgman, Ann, and Rudy Djabbaroff. *Skystars*. New York: Atheneum, 1981. For young adults. Includes a section on Cochran.

Keil, Sally Van Wagenen. *Those Wonderful Women in Their Flying Machines. New York: Rawson, Wade, 1979. Story of the Women's Airforce Service Pilots (WASPs).*

Lomax, Judy. *Women of the Air*. New York: Ballantine/Ivy Books, 1988. Includes a chapter on Cochran.

Moolman, Valerie. *Women Aloft*. New York: Time-Life Books, 1981. Includes extensive information on Cochran.

Smith, Elizabeth S. *Coming out Right*. New York: Walker & Co., 1991. Biography of Cochran for young adults.

Taves, Isabella. "Lady in a Jet." *Reader's Digest,* August 1955. Provides overview of Cochran's life and career; written soon after Cochran became the first woman to break the sound barrier.

Geraldine Fredritz Mock
(1925–   )

In 1964, Jerrie Mock, who called herself a "flying housewife," became the first woman to fly alone around the world.
(Courtesy National Air and Space Museum, Smithsonian Institution, SI Neg No. 77-7948)

---

*T*he dignified Egyptian official, dressed all in black, "would have done justice to any undertaker's parlor," as Jerrie Mock wrote later. He was also in no more hurry than a funeral director would have been.

Mock had already paid the man the required fees for landing her private plane at the Cairo airport. She was anxious to begin preparing for the next leg of her round-the-world trip. The official, however, showed no signs of letting her go. Instead he chatted calmly with Peter Barker, a member of the American Embassy who was helping Mock in Cairo.

After a while, the official picked up a telephone on his desk and tried to make a call. The phone seemed not to be working. Five minutes later, he again picked up the receiver and listened. This time he hung up without even dialing. Then he took an old hand-crank telephone out of a desk drawer and tried that. He alternated between phones for over an hour. Finally, he called in a messenger and sent a note instead—to someone one floor up in the same building!

This experience was typical of many that Jerrie Mock had in March and April 1964, when she became the first woman to fly a plane alone around the world. Mock encountered a few frightening moments in the air, but most of her difficult times happened on the ground.

Born in 1925 in Newark, Ohio, Geraldine Fredritz took her first plane ride when she was seven. She told her fourth-grade class that she would "fly airplanes" when she grew up. She even insisted she would fly around the world.

Jerrie Fredritz continued her interest in aviation by majoring in aeronautical engineering at Ohio State University. So did Russell C. Mock, whom Jerrie had met in her high school algebra class. She and Mock married on March 21, 1945. They set up a home in Bexley, a suburb of Columbus, Ohio, and had three children, Roger, Gary, and Valerie.

Russ Mock went to work in advertising. Like George Palmer Putnam, Amelia Earhart's husband, he became an expert in publicity. Some of his promotional ideas wakened new interests in Jerrie. In 1951, for example, he persuaded her to host a local television show in which high school students discussed parents, politics, and other topics. At first she had stage fright, but then she found she enjoyed the work. She later worked on several other radio and TV programs.

The Mocks also continued their interest in flying. They obtained their private pilot's licenses in 1958 and bought two small planes. One was a Cessna 180 with the federal registration number N1538C. Jerrie called the little plane Charlie, after the code word for the letter *C* in the international alphabet aviators use.

According to *Newsweek* magazine, the idea for Jerrie Mock's record-setting flight grew out of a conversation that sounds like part of a television comedy. "If I don't get out of this house, I'll go nuts," Mock is supposed to have said one evening in 1962.

"Why don't you get in the plane and fly it somewhere?" Russ Mock replied.

"Like where?"

"Like around the world."

Russ may have meant his suggestion as a joke at first. After all, Amelia Earhart had died trying to become the first woman to fly a plane around the world. Earhart had been an experienced professional pilot, aided by an equally experienced navigator and flying a new twin-engine plane large enough to seat 12 people. No woman since that time had succeeded in doing what Earhart attempted. For Jerrie Mock to try it with a mere 900 hours of flight time, a private pilot's license, and an 11-year-old, single- engine plane seemed about as sensible as planning to drive across the United States in a golf cart.

The more the Mocks thought about the idea, however, the better they liked it. Jerrie wanted to sightsee around the world. Russ, true to his advertising background, jumped at the chance to put his wife's name in aviation record books.

The Columbus *Dispatch* agreed to pay most of the trip's costs. In return, Jerrie promised to send the newspaper exclusive stories throughout her journey. Twenty-four companies, most of which made aircraft or aircraft parts, also contributed money or equipment.

In preparation for the long trip, the Mocks replaced three of Charlie's four seats with extra fuel tanks. They gave the plane

Before Jerrie Mock's round-the-world trip, her Cessna 180 was officially named the Spirit of Columbus *to honor her home city of Columbus. To Mock, however, the little plane would always be "Charlie."*
(Courtesy National Air and Space Museum, Smithsonian Institution, SI Neg No. 92-8000)

a more powerful engine and additional navigation equipment. Finally, it received a shiny coat of red and white paint and a new name, *The Spirit of Columbus.* (Charles Lindbergh's plane had been named *The Spirit of St. Louis.*) To Jerrie Mock, however, her plane would always be Charlie.

Jerrie had hoped for a leisurely journey, but in January 1964, a few months before she planned to take off, the trip became a race. Another American pilot, 27-year-old Joan Merriam

Smith, announced that she, too, was going to try to fly around the world. Smith was a professional pilot who had logged 8,200 flight hours. She was flying a twin-engine Piper Apache, a somewhat larger and faster plane than Mock's.

Smith got a head start in the race by leaving from Oakland, California, on March 17, 1964. Jerrie Mock and Charlie took off from Columbus, Ohio, two days later. Dizzied by the shouting crowds and popping flashbulbs at the Columbus airport, Mock was not sure she would even get off the ground. Somehow, however, her hand pressed the starter button. Then, as she wrote in *Three-Eight Charlie,* her book about her trip, "The tiny plane raced down the runway and . . . leaped into the air, eager to explore the world."

Mock's first stop was Bermuda, a British-owned island in the Atlantic. She was delayed there for almost a week, waiting for safe weather and the repair of one of her radios. Luckily for her, Joan Smith was facing similar delays.

Mock finally left Bermuda and headed for the Azores, islands off the west coast of Africa. During the first part of her night flight, she was thrilled by the beauty of the starlit sky. Soon, however, storm clouds surrounded her. She had passed her test for instrument flying shortly before she left, but this was the first time she had needed this skill.

Like Amelia Earhart during her solo transatlantic flight, Mock found that ice was forming on her plane's wings and dragging it down. Unlike Earhart, however, Mock had to radio the nearest air traffic control tower—the one in the Azores—and then wait to be given permission to fly above the clouds. She got wing ice again when she flew from the Azores to Casablanca, in the North African country of Morocco.

Casablanca means "white house." The city had been named for its white buildings, which shone in the African sun. A famous movie had been set in Casablanca, and to Mock its name meant romance. As she was to do whenever weather or other problems grounded her, she used her enforced spare time to tour the city.

Mock had intended to fly next to Tunis, but continuing bad weather forced her to go instead to Boné, a small city in Algeria. The threat of sandstorms then diverted her to Tripoli, in Libya.

Soon after she was airborne on her way to Tripoli, Mock noticed that the long antenna for her high-frequency radio had unreeled. She turned on the electric motor that would pull the antenna wire back in—and forgot about it. Then, when she was over the Sahara Desert, she began to smell something burning. The little motor had long since reeled in the antenna and, pulling uselessly, was slowly burning itself up. She turned it off, but she saw that the still-smoking motor was just inches from one of her gas tanks. If the motor heated the tank enough, the plane would explode.

Panic almost took over. Mock had no parachute, so she could not jump out. Even if she could have, she would not have survived long in the desert. Trying to land might trigger the very explosion she feared. She did not dare try to unwind the antenna again, and without it no one could hear her radio calls for help. She finally realized that all she could do was let the hot metal cool—and pray. Fortunately, that proved to be enough.

After Tripoli, Mock planned to go to Cairo, Egypt's capital city. Finding Cairo seemed easy enough; all she had to do was home in on the powerful radio signal from the navigation beacon near the Cairo airport. She wondered, however, why the air controller in Cairo was still asking for her position after she had landed. Puzzlement turned to alarm when three trucks full of armed soldiers roared down an adjoining runway and surrounded Charlie. Mock learned that she had mistakenly landed at Inchas, a secret air force base just outside Cairo. After many telephone calls to Cairo, the soldiers finally let her go.

Being allowed to go on to Cairo by no means ended Mock's troubles. Indeed, what she called "the obstacle course of red tape" reached new heights in the Egyptian capital. As early as Casablanca, she had begun to learn that flying in Africa was "a little more difficult" than flying in the United States. She later ran into similar problems in the Middle East and Asia. Before she started her trip, she had obtained permission to fly over and land in all the countries she thought she might visit, but that did not seem to help much. Each country did things a little differently, and all—by American standards, at least—did them very slowly.

Part of the problem, Mock realized, was that most aviation in these countries was either commercial or military. No one knew what to make of a self-described "flying housewife" in a tiny private plane. She ran into the obstacle course not only at airports but when phoning Russ and sending her stories to the *Dispatch*, both of which she tried to do at every stop.

The only things that saved Mock's sanity, besides her own persistence, were the assorted friends of friends, aviation officials, and even complete strangers who came to her aid. For example, Peter Barker, the embassy official who helped her in Cairo, was an acquaintance of friends of the Mocks who had once worked in Egypt. People like Barker not only helped Mock navigate bureaucratic mazes but gave her food and places to stay. They even drove her to the airport at 3:30 A.M., the time at which she usually began her departure preparations.

After Cairo, Mock followed the Trans-Arabian oil pipeline to Dhahran, Saudi Arabia. She then flew across the Persian Gulf to Karachi, in West Pakistan. Continuing on to the Indian cities of Delhi and Calcutta, she saw many contrasts. In the countryside, "mile after mile of parched farms" waited for the rain of the monsoon (storm) season. Most of the people she saw in Delhi were "clean, carefree, and gaily attired." In overcrowded Calcutta, however, ragged families lived in the streets and slept on the sidewalks.

In Bangkok, Thailand, Mock encountered the Asian version of the bureaucratic obstacle course. It proved just as annoying as the African and Middle Eastern ones. Trying to make up lost time, she decided to fly directly to Manila instead of refueling in Southeast Asia. This long flight over the South China Sea used nearly all her gas. She was glad to reach the Philippine capital and leave Charlie at a Cessna repair shop, practically the only such shop she found during her trip, for refueling and a complete overhaul.

After Manila, Mock faced several long flights across the Pacific Ocean. Failure to find an island refueling stop in the Pacific had doomed Earhart, but the Pacific flights proved to be among the easiest on Mock's journey. She wrote that on her way to Hawaii, after stopping at the islands of Guam and Wake, she

felt like a queen. . . . My subjects, the foamy clouds and glowing rainbows, put on a command performance just for me. It was worth all the hard work, worry, and sleepless nights. . . . After Christopher Columbus discovered America, he became the Admiral of the Ocean Seas. In my red-and-white Spirit of Columbus, *. . . I became Queen of the Ocean Skies.*

Part of Mock's happiness came from the fact that she was "heading home." All her stops after the Philippines were in land controlled by the United States. Furthermore, a phone call from Russ told her that Joan Smith was now too far behind schedule to catch up. (Smith's journey ultimately took almost twice as long as Mock's.)

Mock arrived in Honolulu full of energy and ready to see Hawaii. She discovered, however, that Russ had telephoned to cancel several parties that had been planned for her because, he said, she needed her sleep. "How could you ruin things before I even got here?" she demanded when she called him back. She knew Russ was just trying to help, but the two had been working at cross purposes for most of the trip. Many of his phone calls, fueled by concern about Joan Smith, had been demands that Jerrie hurry. His inability to understand the frustrating situations she was encountering had irritated her so much that she had hung up on him more than once.

After the longest nonstop flight of her trip—2,410 miles—Mock reached Oakland, California, on April 16. Her first landing in the continental United States drew the second-largest crowd that Oakland Airport had ever held. Russ was there to greet her. He reported that four-year-old Valerie had said to "tell mommy I love her whole lots, as much as the whole world, and the ocean, and the sky."

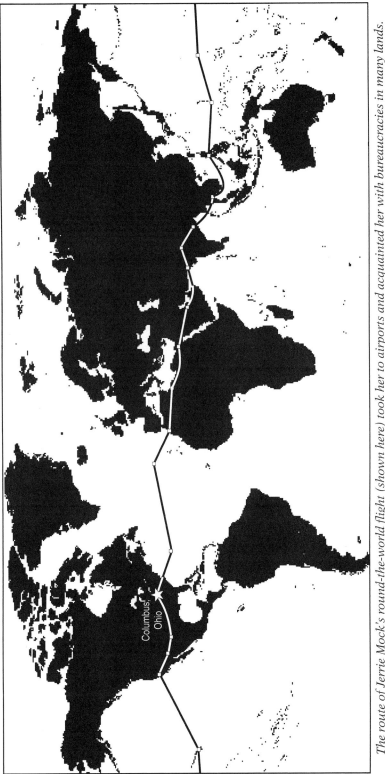

The route of Jerrie Mock's round-the-world flight (shown here) took her to airports and acquainted her with bureaucracies in many lands.

(Courtesy Katherine MacFarlane)

Columbus, Ohio

Flying toward Columbus on April 17 after stops in Tucson and El Paso, Mock reflected on her journey with Charlie. She had come to think of the little plane as "almost alive—a real friend. . . . I patted the top of his tan instrument panel and thanked him for taking such good care of me."

The crowd waiting for Mock in Columbus was even bigger than the one that had seen her off 29 days and 22,859 miles earlier. Everyone from the governor of Ohio on down was there to welcome her. Both Joan Smith and President Lyndon B. Johnson sent telegrams of congratulations. The speeches and cheering put the exhausted Mock "in a daze." She was grateful to return home at last and go to sleep.

Mock won several awards for her flight, including the Federal Aviation Agency's decoration for exceptional service, the country's highest civil aviation award. President Johnson presented the award, a gold medal, on May 4. Mock also received the Ohio Governor's Award, the Amelia Earhart Memorial Award, and the National Pilots' Association Pilot of the Year Award. Charlie was added to the Smithsonian National Air and Space Museum's collection of famous planes.

During and after her trip, Jerrie Mock set a total of 21 speed and distance records. Many of her records still stand. She gave up flying in 1969, however, because her hobby had become too expensive. Her last flights were for missionaries in New Guinea, much like Jerrie Cobb's work in South America.

Jerrie and Russ Mock were divorced in 1979. During the early 1980s Jerrie lived a "gypsy life," traveling around the country in an aging Chevrolet sedan. She stayed with friends or family members or camped out in a tent with her dog, Zappa. She was short of money but enjoyed the freedom to "go where and when I want to."

By 1989, Mock was living with a son in Columbus. She still loved to travel, however. "I don't know if I'll ever settle down," she told an interviewer. "I don't know where I'll be a year from now . . . and I don't want to know."

Chronology

1925	Geraldine Fredritz born in Newark, Ohio
March 21, 1945	marries Russell C. Mock
1951	hosts television show
1958	obtains private pilot's license
1962	begins planning round-the-world trip
January 1964	hears that Joan Smith will attempt similar trip
March 19, 1964	leaves Columbus, Ohio, on round-the-world trip
March 26, 1964	encounters wing ice between Bermuda and Azores
March 29, 1964	tours Casablanca
March 31, 1964	burns out antenna motor on way to Tripoli
April 1, 1964	lands by mistake at Inchas air force base
April 2, 1964	encounters bureaucrats in Cairo
April 5–7, 1964	tours India
April 9–10, 1964	repairs Charlie in Manila
April 13, 1964	flies across Pacific to Hawaii
April 16, 1964	reaches Oakland, California
April 17, 1964	returns to Columbus
May 4, 1964	receives medal from President Johnson
1969	retires from flying
1979	divorced from Russell Mock
1980s	leads "gypsy life"

Further Reading

Genett, Ann. *Contributions of Women: Aviation.* Minneapolis, Minn.: Dillon Press, 1975. For young adults. Contains a chapter on Mock.

Mock, Jerrie. *Three-Eight Charlie.* Philadelphia: Lippincott, 1970. Mock's account of her round-the-world flight.

"Shades of Amelia." *Newsweek,* March 30, 1964. Describes plans for Mock's trip and the race with Joan Smith.

Geraldyn Cobb
(1931–)

In 1960, Jerrie Cobb passed the tests given to male astronaut
candidates, but NASA would not accept her or any other women
into the astronaut program.
(Courtesy National Air and Space Museum, Smithsonian Institution,
SI Neg No. 79-6360)

*O*n a September morning in 1959, Jerrie Cobb and her employer, Tom Harris, met two men on the beach in Miami, Florida. When Harris introduced the men, Cobb recognized their names immediately. Dr. Randolph Lovelace and Brigadier General Donald Flickinger were experts in air and space medicine. They advised the National Aeronautics and Space Agency (NASA) on Project Mercury, the first U.S. manned space program.

The men did not recognize Cobb's name. The blonde, ponytailed woman looked younger than her age of 28. Lovelace and Flickinger were amazed to learn that she had logged more than 7,000 hours of flight time during a 16-year career in the air. She had also set three aviation records.

Amazed—and very interested. Rumors suggested that the Soviet Union was planning to put a woman into space, the men explained. They, and NASA too, wanted to know whether women pilots could pass the demanding physical and psychological tests that the Mercury astronauts had taken. If women could pass the tests, the astronaut program might be opened to them. "Would you be willing to be a test subject for the first research on women as astronauts?" the men asked Cobb.

Jerrie Cobb could hardly believe her luck. All her life she had loved flying. How could she refuse a chance to fly higher than any woman had gone before?

━━━━━━━━━━

The idea of a human being going into space was strictly science fiction when Geraldyn Cobb was born in Norman, Oklahoma on March 5, 1931. Her father, William Harvey Cobb, sold cars when he could, but sales were few in those early years of the depression. Geraldyn's mother, Helena, planned to be a teacher. They already had one daughter, Carolyn.

Jerrie, as she was always called, was very different from her older sister. Carolyn liked to play with dolls, but Jerrie would rather ride horses. Carolyn loved people, but Jerrie preferred to be alone. Carolyn chattered, but a speech impediment made talking hard for Jerrie. Even after surgery cured this speech problem, Jerrie's shyness remained.

Harvey Cobb learned to fly at the beginning of World War II. He even bought a used plane—and took his family up in it. Helena and Carolyn were not impressed, but 12-year-old Jerrie was. "Even before [we] . . . had reached 300 feet, I recognized that the sky would be my home," Cobb recalled in her autobiography, *Woman into Space.* "I tumbled out of the airplane with stars in my eyes."

Jerrie immediately began demanding that her father teach her to fly, and he finally gave in. He had to put several pillows on the seat of the old plane so she could see over the edge of the open cockpit. Then he added wooden blocks to the rudder pedals so she could reach them. Those lessons ended, however, when the Cobbs moved and sold the plane.

Jerrie made her way into the sky again when she was 15. By then she was a student at Oklahoma City's Classen High School. The school's football coach was a flight instructor and owned a plane. He gave Jerrie the rest of the training she needed. She got her pilot's license—what she called her "ticket to the sky"—in 1948, on her 17th birthday.

Jerrie graduated from Classen High that June. Instead of going to college, however, she wanted to work for her commercial pilot's license. To earn money for more lessons and flight time, she played three seasons with a semiprofessional women's softball team, the Sooner Queens. By the end of that time, she had both her commercial license and her own plane.

Cobb's first paid flying job, which she got in 1950, was flying "low and slow" over oil pipelines to look for leaks. She quickly learned to recognize telltale smells or stains on plants and soil. Then she got an instructor's license and taught oilfield workers how to fly. Some of her students had doubts about taking lessons from a woman. The formerly shy Cobb, however, found she could "breathe fire" at these tough men and make them respect her.

In 1953, having heard that there were flying jobs in Miami, Cobb moved to Florida. There she found work with an aircraft ferry service named Fleetway, which specialized in delivering small planes to foreign countries. Fleetway's owner, Jack Ford, at first refused to believe that any woman could survive the rough life of a ferry pilot. Still, Cobb persuaded him to let her try.

"Rough life" proved to be an understatement. Cobb spent part of her first ferry trip in jail. En route to delivering a plane to Peru, she stopped in Guayaquil, Ecuador, for refueling. This was just what Ford had told her to do, but she found herself surrounded by armed soldiers on the runway. Ecuador and Peru, it seemed, were having a little "disagreement" at the moment. The Ecuadorean government was not pleased to see a plane carrying the markings of the Peruvian Air Force. Cobb spent 12 days in prison, answering questions and improving her Spanish, before the United States arranged for her release.

For the next two years, Cobb "bounded like a tennis ball between North and South America." At the time, she was the only female international ferry pilot in the United States. She grew used to spending long hours alone in the air, then finding her way around unfamiliar towns on the ground. After one forced landing, she rode back to the nearest city on a banana boat. She prayed that the huge tarantulas falling now and then from overhanging trees would not end up either down her back or in her dinner.

Cobb and Ford gradually fell in love, and for a while they considered marriage. But Cobb came to realize that "to marry Jack would be to cage an eagle." Sadly, she ended her relationship with Ford and left Fleetway in the fall of 1955.

Cobb had gained some experience as a test pilot while at Fleetway. Like Jackie Cochran, she found she enjoyed this work. Trying to set air records was one way to test planes, so she looked for a chance to make such an attempt. Oklahoma was preparing to celebrate 50 years of statehood, and a reporter friend of Cobb's persuaded the celebration committee that having Cobb set a record in an Oklahoma-built plane would draw national attention to the state. The committee agreed to pay Cobb's expenses. Aero Design and Engineering Corporation of Oklahoma City loaned her an Aero Commander, a type of plane she had learned to fly well during her ferrying days.

Cobb first flew from Guatemala City, Guatemala, to Oklahoma City on May 25, 1957 to set a nonstop distance record for the Aero Commander's class (size and weight) of plane. She covered the 1,504-mile distance in a little over 8 hours. Three weeks later, she set an altitude record as well by reaching 30,560 feet, almost six miles up. This flight was a magical experience for her. "I saw the bluest sky I'd ever known. . . . There was no horizon, no boundary. . . . I felt that I could reach up to the sun, or touch the stars that were hidden in its glow."

Unfortunately, Cobb's success worked against her. At the time she set her records, she was working for a company that sold mostly Cessnas. The company was not happy to have her name linked with Aero Commanders. The logical thing, of course, would be for Cobb to work for Aero Design. It took Cobb more than a year, however—and a third record, this time for speed—to persuade Thomas Harris, the company's sales manager, to hire her. She set the speed record on April 13, 1959. Harris agreed to take her on while she tried for the record, then kept her on the payroll afterward.

Cobb accompanied Harris to a meeting of the Air Force Association in Miami in September 1959, where they chanced to meet Lovelace and Flickinger. (Lovelace's early career had been given a boost by another woman pilot, Jackie Cochran.) As a result of their conversation, Cobb found herself reporting to the Lovelace Foundation in Albuquerque, New Mexico, on February 14, 1960, to begin the Mercury astronaut tests. She was understandably nervous, knowing that far more than her own future depended on the results. "Here was the chance, perhaps the only one, to prove a female spaceworthy," she later remembered thinking.

Cobb spent eight or more hours a day at the center for the next five days, taking 75 different tests. On one day, for example, she pedaled an exercise bicycle to the point of exhaustion and beyond. Machines monitored her pulse, blood pressure, and the amount of oxygen she breathed. On another day, to test her heart and circulation, she lay on a table that was moved from a flat to a tilted position. Even strong, healthy pilots sometimes fainted or became dizzy when the position of their bodies was changed suddenly like this because too little blood reached their brains.

She passed every one of the tests. Dr. Secrest, the man in charge of her testing, told her she was "a remarkable physical specimen."

Dr. Lovelace described Cobb's tests at a scientific meeting in Sweden in August. The results suggested, he said, that women might be better suited for space travel than men. "Women have lower body mass, need significantly less oxygen and less food, and may be able to go up in lighter capsules, or exist longer than men on the same supplies," he explained.

Cobb and the Mercury astronauts were given such demanding physical tests because doctors of the time knew little about the conditions an astronaut might face in space or during trips to and from orbit. They felt that a space traveler must be ready for anything. For the same reason, the astronauts had been given a complete range of psychological tests. In September 1960, Cobb went through these same tests, plus another that the astronauts had not yet faced.

In this last test, Cobb floated in a tank of water enclosed by steel walls. The water, warmed to body temperature, made her almost weightless, as an astronaut would be in space. The underground tank room was silent and dark, completely cut off from the outside world. Hidden cameras and microphones recorded Cobb's reactions. She was told to stay in the tank as long as she was comfortable there, but she could ask to leave at any time.

The water tank test measured Cobb's ability to spend long periods alone without any information reaching her senses. Some people under these conditions saw visions or heard imaginary sounds, such as dogs barking. Others found the experience so frightening that they demanded to be let out of the tank after a short time. Cobb, however, felt completely relaxed as she floated in the tank.

Cobb had entered the tank at 9:20 in the morning. When she finally asked to come out, it was 7 P.M. She had stayed in the tank for nine hours and 40 minutes. She learned later that the longest time anyone else had stayed in the tank was six and a half hours. Dr. Shurley, who had carried out the test, wrote, "Probably not one in 1,000 persons would be capable of making such a lengthy isolation run." He told Cobb that she "excel[led] in loneliness."

In May 1961, Cobb went through a third round of testing at the navy's School of Aviation Medicine in Pensacola, Florida. Meanwhile, Dr. Lovelace in Albuquerque was giving the first group of tests to 22 other experienced women pilots. Twelve of the women passed the tests. Two also took the psychological tests, including the tank test, and did extremely well.

But in August, just before the women were to go to Pensacola to take the navy tests, Lovelace told Cobb that the tests had been cancelled. The navy refused to carry out any more testing without orders from NASA, and the orders had not come.

Cobb flew to Washington and saw all the officials she could. As when she had taught the oil workers, she "breathed fire" as she tried to save the program she had come to care deeply about. Jane Hart, wife of Senator Philip Hart of Michigan and one of the 12 women pilots who had passed the astronaut tests, helped Cobb win the chance to have the women's testing discussed in a hearing before a Congressional committee. The hearing was held on July 17, 1962. During it, Cobb repeated the findings suggesting that qualified women could do at least as well in space as men.

Several committee members expressed support for an astronaut testing or training program that involved women, but they were unable to make NASA change its mind. The United States did not accept female astronaut candidates until 1978, almost 20 years after Jerrie Cobb was tested. Sally Ride, the first American woman to go into space, went on her space shuttle mission in 1982.

Even though Jerrie Cobb did not achieve her dream of going into space, she won many awards during her aviation career. They included gold wings from the Fédération Aéronautique Internationale, the Amelia Earhart Gold Medal from the Ninety-Nines, the "Woman of the Year in Aviation" award for 1959 from the Women's National Aeronautic Association, and the "Pilot of the Year" award in 1960 from the National Pilots' Association.

Cobb once wrote, "I believe that . . . space exploration will reveal God's creations and purposes more clearly to us." When that path to what she saw as God's purposes was closed to her, she turned to another. In 1964, she went back to South America and became a missionary pilot. Today, calling herself the Amazonas Airlift Service, she flies doctors, missionaries, and medicines to remote villages in the rain forests surrounding the Amazon River. She also carries sick villagers to cities for expert medical care. She has rescued other pilots that have been forced down in the rain forest as well.

Cobb won the Harmon Trophy for the year's best woman pilot in 1973 for her missionary work. She was even nominated for the Nobel Peace Prize in 1981. Oklahoma Representative Mickey Edwards, who proposed her name, wrote that she had "devoted all of her skills and resources to providing health, bringing hope, and creating peace for thousands of men, women and children." Cobb is happy in her work, perhaps even happier than she would have been in space. "It's a privilege to do the kind of work I do," she has said. "It's more fulfilling than anything I can think of."

Chronology

March 5, 1931	Geraldyn Cobb born in Norman, Oklahoma
1943	takes first plane ride and flying lessons
March 5, 1948	earns private pilot's license
1950	takes first paid flying job
1953	begins working as Fleetway ferry pilot
1955	leaves Fleetway
May 25, 1957	sets distance record in Aero Commander
April 13, 1959	sets speed record; hired by Aero Design
September 1959	meets Dr. Lovelace and Brigadier General Flickinger, aerospace experts
February 14, 1960	begins taking Mercury astronaut tests
February 28, 1960	passes physical tests
August 1960	Lovelace presents results of Cobb's tests to science meeting
September 1960	Cobb passes water tank test and other psychological tests
May 1961	takes navy tests in Florida; testingof other women pilots begins
August 1961	testing of women pilots cancelled
July 17, 1962	congressional hearing on astronaut testing of women

1964	Cobb becomes missionary pilot in Amazon
1973	wins Harmon Trophy
1981	nominated for Nobel Peace Prize
1990s	still flying missionary runs as Amazonas Airlift Service

Further Reading

Briggs, Carole S. *At the Controls: Women in Aviation.* Minneapolis, Minn.: Lerner, 1991. For young adults. Includes a chapter on Cobb.

Cobb, Jerrie, with Jane Rieker. *Woman into Space.* Englewood Cliffs, N.J.: Prentice-Hall, 1963. Cobb's autobiography; describes her astronaut testing in detail.

"Cobb, Jerrie." *Current Biography* 1961. Summarizes Cobb's career and briefly describes her astronaut testing.

Genett, Ann. *Contributions of Women: Aviation.* Minneapolis, Minn.: Dillon Press, 1975. For young adults. Includes a chapter on Cobb.

Hodgman, Ann, and Rudy Djabbaroff. *Sky Stars.* New York: Atheneum, 1981. Includes information on Cobb.

"A Lady Proves She's Fit for Space Flight." *Life,* August 29, 1960. Describes the physical tests Cobb passed.

Rieker, Jane. "Up and Up Goes Jerrie Cobb. *Sports Illustrated, August 29, 1960. Describes Cobb's life and the first phase of her astronaut testing.*

Bonnie Linda Tiburzi
(1948–   )

*Bonnie Tiburzi, the first woman to be hired as a pilot
by a major U.S. airline, achieved her dream of
becoming a captain (pilot in charge of a flight crew) for
American Airlines in 1988*
(Courtesy Bonnie Tiburzi)

*T*he woman passenger boarded the front of the American
Airlines Boeing 727. She glanced briefly into the jet's cockpit,
where the pilot, copilot, and flight engineer were preparing for
takeoff. Then her face took on an expression of horror, as if she
had seen an unpleasant insect. *"What is that?"* she demanded.

The woman was staring at the flight engineer. The engineer, dressed in uniform, was quietly carrying out the duties that went with the job. The only unusual thing was that the engineer was also a woman. She was Bonnie Tiburzi, the first woman pilot hired by a major U.S. airline.

After starting to fly for American in 1973, Tiburzi encountered few reactions against her that were as strong as this. Once she showed that she could do her job, most of the pilots she worked with respected her and treated her as one of themselves. Most passengers, too, greeted the sight of a woman in the cockpit with surprise, perhaps, but no alarm. But before being allowed to climb into a 727 cockpit, Tiburzi had to overcome the almost universal view that, no matter how many women might fly and even set records in small planes, flying big commercial aircraft was a man's job.

———

Bonnie Linda Tiburzi was born on August 31, 1948. She spent her childhood in Ridgefield, Connecticut. Like the Stinsons, the Tiburzis were a flying family. August Robert ("Gus") Tiburzi, Bonnie's Italian father, owned a small business called Tiburzi Airways. Based at nearby Danbury irport, the business included a flying school and planes that could be hired for charter flights. (This "fixed-base operation," as pilots call it, was similar to the many airport businesses that Edna Whyte ran.) Gus Tiburzi had met his future wife, Ingabritt Gunvor, in Sweden, where he had flown for Scandinavian Airlines (SAS). In Danbury, Ingabritt Tiburzi helped her husband run Tiburzi Airways as well as raising three children, Allan, Anita, and Bonnie. Both Allan and Bonnie would later become commercial pilots.

Like Jerrie Cobb, Bonnie Tiburzi took her first flying lesson when she was 12 years old. By then Bonnie's father had taught her the uses of the instruments in half a dozen kinds of planes. She spent every spare minute at the airport, which seemed far more interesting than school.

When Bonnie was about 15, Tiburzi Airways closed. The family moved to Pompano, Florida, where Gus Tiburzi started a travel agency. Deprived of airplanes, Bonnie turned to sports. As Sally Ride would do a little later, she became an excellent tennis player. She also went to the beach and surfed with the boys.

But Bonnie did not forget aviation. When asked what she wanted to do as an adult, she always answered, "I'm going to be an airline pilot, just like my father was." She refused to listen when people told her that this was impossible because there were no American women airline pilots.

Bonnie graduated from high school in 1967, then traveled and worked in Europe for about a year. On her return, she tried college for a few months, but—again like Jerrie Cobb— she realized that the only education she really wanted was flying lessons.

These she could find at the Pompano airfield. But strangely, when her instructor told her she was ready to fly solo, Tiburzi panicked. "For me, the solo was symbolic," she wrote in her autobiography, *Takeoff!* "If I did it, I was committed to a profession whose doors were closed to women." She fled to New York, where her sister lived, and worked there for several months before deciding to return to Pompano and flying. Then, in the summer of 1969, she soloed and "had never been so happy. The only wonder was that it had taken me so long to take the most important step of my life." After eight more months, she won her private pilot's license.

During the next two years, Tiburzi earned her commercial, instructor's, instrument flying, and multiengine pilot's licenses. As Jerrie Cobb had done, she supported herself and gained flying experience with a patchwork of odd jobs. They included ferrying planes, instructing flight students, and flying charters. (One of the oddest jobs was flying Thanksgiving turkeys to construction workers on an island off the Florida coast.) By 1972, however, she felt a growing frustration. She had flown dozens of types of small planes and had put in almost

102

a thousand hours of flight instruction time, but she seemed no nearer to her goal of working for an airline.

On another visit to New York, Tiburzi's sister introduced her to the editor of *Harper's Bazaar,* a glossy national magazine. The editor was planning to do an issue on Florida and asked Tiburzi, who was as slender and beautiful as a model, to pose for pictures in it. She agreed. The story about her, which appeared in January 1973, mentioned her flying ambitions. Tiburzi liked the accompanying photographs but laughed to see herself described as a "jet-setter."

Shortly afterward, Tiburzi got a surprise phone call. Gene Steele, a lawyer for the Federal Aeronautics Administration (FAA) in Washington, D.C., had seen the *Harper's* story. He wanted to offer her a tip: "American's hiring." Tiburzi said she had applied to that airline eight months before without success. Steele told her to update her résumé and try again.

At the end of January, Tiburzi took Steele's advice.

American Airlines' personnel office called her a week later and asked her to come to Dallas, Texas, for an interview and a physical. "As I hung up," Tiburzi wrote later, "my screech of delight sent shock waves across the room."

Tiburzi took her written test and interview on February 20. She did so well that she was told to take the full pilot's physical exam instead of the shorter physical given to most applicants. She then went back to Florida to wait— and try not to think about the fact that only 214 pilots would be chosen out of 15,000 applicants. Her wait ended with a telegram on March 15: American had accepted her as a flight crew member trainee. She was to begin her classes in Dallas on March 30.

When Tiburzi arrived in Dallas, she had to explain to the driver sent to pick up students that she wanted to go to the Inn of the Six Flags, where the pilot trainees stayed— *not* to the Flight Attendant College. She had to wait while the woman in the inn lobby found a separate room for her. Away from her parents and friends, the only woman in the pilots' dormitory, at first she felt terribly lonely.

After American Airlines hired Bonnie Tiburzi as a pilot in 1973, she went through a training program in which she learned how to handle all the controls in a Boeing 727 jet.
(Courtesy American Airlines and Bonnie Tiburzi)

Tiburzi's class at the Flight Academy had 20 members. At age 24, she was the youngest. Most of the others were college graduates. Some were military pilots with extensive flying experience. Tiburzi would have been worried about keeping up with them under any circumstances. Her worry was greater, however, because she felt responsible to more than herself. Her acceptance into pilot training school had been publicized as a step forward for all women. If she failed, she knew it would be a major setback for other women pilots.

The pilot training course was 10 weeks of hard work. Tiburzi and the other trainees studied Boeing 727 jets exhaustively in ground school classes. Then continued their studies in the simulator, which contained all the controls in a 727 cockpit but did not fly. They also learned the duties of the three crew members who flew the big jets: the captain, copilot, and flight engineer. The flight engineer's job was the most important to them because all newly hired pilots were assigned to this junior position.

The captain and copilot took turns flying the plane, Tiburzi learned, while the flight engineer sat behind them. The engineer did the preflight checks of the plane, read out values on the control dials, called out and marked off checklists of procedures, and generally aided the captain and copilot. Tiburzi wrote that the flight engineer's job combined the roles of "technician, second officer, third pilot, odd-jobber and . . . troubleshooter." Astronaut Sally Ride later would have this same job on the space shuttle.

Most of the time, Tiburzi's teachers treated her just like any other pilot trainee. As her classmates saw that she could work as hard and do as well as any of them, they too began to accept her. She ate meals, played sports, and socialized with them. She was no longer lonely.

Finally the 10 weeks were up, and it was time for tests. Tiburzi took the same kinds of exams as her classmates, but she was tested more thoroughly. Most students' oral examinations, for example, took about 45 minutes. The tests covered just a few of the plane's systems, such as the electrical and fuel systems. Tiburzi, however, had to spend two hours describing "everything."

Tiburzi passed her oral exam, simulator check ride, and aircraft check ride. She was now ready to graduate. American had kept the media away from her during her training, but the company scheduled a press conference for June 4, the day she was given her "wings." There Captain Ted Melden, the vice president for flight at the Dallas center, pinned the wings on the new uniform that had been specially designed for Tiburzi. He then kissed her on the cheek.

Tiburzi faced a different kind of test on her first flight into Chicago's O'Hare Airport, where the crew lounges were segregated by sex. What would she do, she wondered, when she saw the lounge sign, "MALE CREW MEMBERS ONLY"? Would she have to leave the captain and copilot and go to the lounge for the female flight attendants?

When Tiburzi reached the lounges, however, she found that her problem had been solved for her. At the bottom of the "MALE CREW MEMBERS ONLY" sign someone had written in bold letters, "AND BONNIE TOO!" Soon after that, American's lounges at all airports were opened to all crew members regardless of sex or position.

Tiburzi found a more important kind of acceptance after her first air emergency. It occurred after she had been working as a flight engineer for only a few weeks. Her jet had just taken off from New York when air suddenly whooshed out of the cockpit. The cabin pressure had dropped abruptly. Tiburzi knew this could mean one of two things. Either there was a hole in the plane, which could lead to a dangerous loss of oxygen in the whole aircraft, or a stuck outflow valve was pulling air out of the cockpit. Tiburzi's control panel gave no clue about the cause of the problem.

Everything now seemed to take place in slow motion. The crew members put on their oxygen masks so they would not black out. Tiburzi began the emergency procedures she had learned in the simulator. Then, however, she heard the captain tell the copilot, "Turn around and help her." As the copilot started to obey, Tiburzi snapped, "Would you please turn around and do your job." The copilot did.

After a moment, the cabin pressure started to return to normal. Tiburzi and the others never found out exactly what had gone wrong, but a jammed outflow valve seemed the most likely guess. In any case, Tiburzi received no more unwanted offers of help. She had shown that she knew how to do her job in an emergency and would not panic. When the plane landed, she felt that "now I was definitely a fully functioning and integrated part of the team."

Being accepted outside the cockpit was sometimes harder. Few passengers reacted as strongly as the woman who said "What is that?" at the sight of Tiburzi in the cockpit. Still, many people did not know quite what to think of her. Passengers often assumed she was a flight attendant. On the ground, her uniform made people take her for a policewoman, a bus driver, or a hotel bellhop. Even mechanics and other airport professionals sometimes questioned her orders. These problems occurred less often, however, as time went on.

Male pilots respected Tiburzi as a coworker, but she sometimes felt lonely because her job seemed to leave no time for romance. This changed late in 1973, however, when she met an "impossibly good-looking" fellow pilot whom she calls Mike in her autobiography. At first Mike seemed perfect for her because, as a fellow pilot, he understood both the joys and the demands of her job. They married on April 20, 1974. But they had difficulty matching their work schedules, and other problems developed as well. They divorced in 1979.

At the end of 1973, soon after Tiburzi met Mike, American furloughed, or laid off, all 214 of the pilots it had hired the previous spring. That included Tiburzi. The pilots had no idea whether they would be rehired. Tiburzi found ground jobs, but she wanted to be back in the air.

American rehired Tiburzi in April 1976. In 1979, six years after she had first been hired, Tiburzi was promoted to copilot. She was thrilled to have a chance to fly one of the big jets for the first time. Unfortunately, she had the experience for only a few months. In 1980, due to a slump in the economy, she and 200 other new copilots were returned to being flight engineers.

Tiburzi realized that once again her work had become her whole life. She decided it was not enough. She began to make a point of meeting other women airline pilots. There were now

Bonnie Tiburzi still loves flying American Airlines's most sophisticated jets, but she wishes she had more time to spend with her husband and children
(Courtesy Bonnie Tiburzi)

enough of them to form a group, the International Social Affiliation of Women Airline Pilots. Tiburzi attended their second annual meeting in 1980, and many of the women she met there became her friends. She was delighted to share "airplane talk—hangar flying . . . larded . . . with [talk about] shades of nailpolish and . . . each other's uniforms."

Tiburzi's private life also took a happier turn. In 1983 she married a New York politician who later became a lawyer. (To protect her family's privacy, she prefers not to give his name.) He knew little about flying but had great confidence in her abilities. They now have two children, Tony and Britt.

Bonnie Tiburzi still flies for American Airlines, where she is one of 200 women pilots. She regained the poisition of copilot and then, in 1988, she achieved her dream of being promoted to captain, the pilot in charge of a jet's crew. She flies Boeing 757s and 767s, American's most sophisticated jets, to international destinations ranging from Europe to South America.

In spite of her success, Tiburzi has found that, like many other career women with families, she wishes she could spend more time with her husband and children. "I would love to retire, but I'm too young," she says. Nonetheless, she still loves flying just as much as when she wrote in her 1984 autobiography, "Flying jets is exhilarating. . . . The airplane is . . . a world above a world—and you are in it . . . looking out and down at miracles."

Chronology

August 31, 1948	Bonnie Tiburzi born in Ridgefield, Connecticut
1961	takes first flying lesson
1967	graduates from high school; travels and works in Europe
1970	earns private pilot's license
January 1973	featured in *Harper's Bazaar* article
February 1973	takes written test, interview, and physical at American Airlines
March 30, 1973	begins pilot training program with American Airlines in Dallas
June 4, 1973	graduates from pilot training program
December 31, 1973	laid off by American Airlines
April 20, 1974	marries Mike
April 1976	rehired by American Airlines
1979	divorced from Mike; becomes copilot
1983	marries New York politician and lawyer
1984	autobiography published
1988	becomes captain for American

Further Reading

"Bonnie Tiburzi: In Control," *Harper's Bazaar,* January 1973. Brief article with photographs that helped to launch Tiburzi's career.

Briggs, Carole S. *At the Controls: Women in Aviation.* Minneapolis, Minn.: Lerner, 1991. For young adults. Includes a chapter on Tiburzi; focuses on her training and early career.

Crateura, Linda B. *Growing Up Italian.* New York: Morrow, 1988. Describes how being brought up as an Italian-American helped shape the personalities and careers of certain famous Americans. Includes a chapter on Tiburzi.

Chadwick, Bruce. "Women in Aviation." *Cosmopolitan,* November 1985. Includes information on Tiburzi and other women airline pilots as well as background on contemporary women pilots.

Holden, Henry M. *Ladybirds.* Includes information on Tiburzi and other women airline pilots.

Tiburzi, Bonnie. *Takeoff!* New York: Crown, 1984. Tiburzi's autobiography.

Sally Kristen Ride
(1951–   )

In 1983, Sally Ride achieved Jerrie Cobb's dream of being the first American woman to go into space. Today, Ride heads an institution devoted to space research.
(Courtesy Sally Ride)

Five . . . *four* . . . *three* . . . *two* . . .

"The rockets light! The shuttle leaps off the launch pad in a cloud of steam and a trail of fire." So Sally Ride described the beginning of her June 1983 space shuttle flight in a children's book, *To Space and Back.* In an interview made a few months after the flight, she added a description of her own feelings: "All of a sudden you know you're *going.* . . . There is nothing like it. . . . It literally *overwhelms* you."

On her historic flight, Ride fulfilled Jerrie Cobb's dream of becoming the first American woman to go into space. She also helped to launch two communication satellites into space and tested a 50-foot-long robot "cherry picker" arm that she had helped to design. Since then she has traveled into space a second time, written a report recommending future goals for the United States space program, and headed an academic institution devoted to space research.

———

Sally Kristen Ride was born on May 26, 1951, in Los Angeles, California. She grew up in Encino, a suburb of Los Angeles. Her father, Dale B. Ride, taught political science at Santa Monica Community College. Joyce, her mother, had been a teacher and was active in volunteer work.

Sally and her younger sister, Karen, grew up with few restrictions. "Mostly we just let them explore," their father said. Ride has added, "I can't remember a single time [my parents] ever told me not to do something I wanted to do."

Even as a child, Sally loved adventure. Science fiction, Nancy Drew stories, and James Bond novels were her favorite reading. She was athletic and adventurous in real life, too. Her sister recalls, "When the kids played baseball or football out in the streets, Sally was always the best. . . . She was the only girl who was acceptable to the boys."

Joyce Ride introduced Sally to tennis when Sally was about 10 years old. Sally eventually became good enough to rank 18th among junior players in the United States. Tennis competition helped her develop the independence, self- confidence, and coolness under pressure she showed so clearly as an astronaut.

Sally's tennis skill won her a scholarship to Westlake School for Girls in Los Angeles. One of her teachers there, Dr. Elizabeth Mommaerts, stirred her interest in science. Ride said later that Mommaerts, who taught physiology (the study of how the body functions), was "logic personified." But the part of science that finally attracted Sally most was physics, the study of matter and energy.

Sally Ride enrolled in Swarthmore College in Pennsylvania in 1968. After three semesters she dropped out, however. Neither the Swarthmore tennis team nor the chilly Pennsylvania climate was to her liking. Ride returned to southern California, where she spent most of her time on tennis. Tennis star Billie Jean King played a match with her and recommended that she take up the sport professionally.

Ride transferred to Stanford University, in northern California, in 1970. She continued to play tennis for a while but finally quit because her skill could not quite match her will to excel. "It was obvious that I wasn't going to be one of the top players, so I gave it up." Ride's mother says Sally stopped playing tennis because "she couldn't make the ball go just where she wanted it to."

Ride majored in physics at Stanford, but her roommate, fellow tennis player Molly Tyson, interested her in English as well. Ride stretched the usual four-year college term to five years and completed a double major, in physics and English. In 1973, she graduated with honors in both.

Finally choosing science over English, Ride did graduate work at Stanford in astrophysics, the physics of stars and other objects in deep space. One day in 1977, while looking for a job to follow graduate school, she saw an advertisement in the campus newspaper. It urged young scientists to apply to NASA for positions as mission specialist astronauts on future space shuttle flights. The space shuttle, which had not yet flown, was a new kind of spacecraft that could be reused. NASA hoped it would act as a "space bus," carrying astronauts to Earth's orbit and back several times a year. Ride had never thought about becoming an astronaut before. Still, she says, "I was on my way out of the room to apply while I was still reading the notice in the paper!"

NASA's requirements for astronauts had changed since Jerrie Cobb's day. For example, the agency no longer demanded that all astronauts be military test pilots. As George W. S. Abbey, NASA's director of flight operations, stated, "the pilot's job [was] no longer the prime job" in space flights. Experienced pilots were still needed to fly the shuttle, but NASA now wanted most of its astronauts to be scientists who could carry out experiments while the shuttle was in orbit. These scientists would be the mission specialists. Some of them could be women.

Ride was one of over 8,000 people, including about 1,000 women, who answered NASA's astronaut advertisement. She was among 208 finalists who went to the Lyndon B. Johnson Space Center in Houston, Texas, for testing. She spoke with two psychiatrists and then was interviewed by the astronaut selection committee. She also took extensive medical exams and some tests of physical fitness. She did not have to go through the torturous tests that Jerrie Cobb had passed, however. Aerospace experts did not expect astronauts to face extreme conditions on space shuttle flights.

In January 1978, Ride learned that she was among the 35 astronauts the space agency had selected, which included six women. NASA never explained exactly how it made its choices. However, George Abbey said the agency had looked for people who were self-reliant yet were also "team players."

In the summer of 1978, shortly after getting her Ph.D. degree from Stanford, Ride returned to the Johnson Space Center to begin her astronaut training. The ASCANS (astronaut candidates), as NASA called the first-year students, spent most of their time in classes. They also learned parachute jumping, water survival techniques (in case the shuttle made a forced landing in the ocean), and how to fly a jet.

Learning to fly a plane proved easy for the adventurous Ride. Fellow astronaut trainee and experienced pilot John Fabian, who later would be a crew member on Ride's shuttle flight, said of her, "She's very cool. . . . She understands the airplane, she's a superb co-pilot and . . . an excellent pilot. We'll get out over

Women were allowed to become astronaut candidates for the first time in 1978. Sally Ride was one of six women whom NASA accepted. Here the six women, in the clothing they would wear on their space shuttle flights, pose with a space suit used for space walks during the Apollo program about 10 years before. (Left to right) Shannon W. Lucid, Margaret R. (Rhea) Seddon, Kathryn D. Sullivan, Judith A. Resnick, Anna L. Fisher, and Sally K. Ride.
(Courtesy National Aeronautics and Space Administration)

the Gulf [of Mexico] and start rolling the airplane upside down . . . and . . . she'll say, from the back of the airplane, 'You just don't enjoy this enough.'" Ride found she enjoyed flying so much that she went on to earn a private pilot's license.

The following year, Ride and the other trainees spent most of their time in the simulator, a mock-up of the shuttle craft. They endlessly rehearsed all parts of a shuttle mission, including everything that might go wrong. They also took rides in a jet that let them experience the increased pressure they would feel during the acceleration of the shuttle's takeoff and, by contrast, the weightlessness of space. At the top of the jet's arc, the trainees floated free of gravity for about 30 seconds.

Ride adjusted to weightlessness quickly. She learned to push off gently from walls when she wanted to move and to anchor herself to fixed objects when she wanted to stay still. She later wrote, "The best part of being in space is being weightless. It feels wonderful to be able to float without effort; to slither up, down, and around the inside of the shuttle just like a seal."

Weightlessness was not the only joy that Sally Ride discovered during her astronaut training. She became close friends with one of the other trainees, a tall, thin, red- haired astronomer named Steven Hawley. Hawley and Ride were married on July 24, 1982. Ride flew her own plane from Texas to their wedding in Kansas.

While waiting for their turn in space, the astronauts worked on projects on the ground. Ride's project was a cranelike mechanical arm. It would be used to release and retrieve satellites and experiment packages outside the shuttle. The arm, called the Remote Manipulator System (RMS), was designed in Toronto, Canada. Ride became one of the two astronauts most skilled in using it (John Fabian was the other). She even suggested improvements in its design.

Ride also served as a CAPCOM (capsule communicator) during the second and third flights of the space shuttle, which took place in November 1981 and March 1982. She was the first woman to hold this post. The CAPCOM relays instructions from the flight director in Mission Control to the shuttle crew in space. An astronaut acting as CAPCOM needs a detailed knowledge of the shuttle flight process, coolness under pressure, and the ability to speak concisely and clearly. (Not only the shuttle astronauts but a large television audience hear the CAPCOM's voice.) Astronauts chosen as CAPCOMs were likely to be selected for future space missions.

Navy Captain Robert L. Crippen, who had flown on the first shuttle mission in April 1981, was scheduled to be commander of the seventh mission, STS-7. In the spring of 1982, he announced his choice of crew members for the mission. Ride was among them. Crippen said he chose her partly because of her experience with the robot arm, which would be tested in space for the first time during that mission. But he also said that he chose her because she was "easy to be with" and calm under stress. "Sally can get everything she knows together and bring it to bear where you need it." Besides Crippen, Ride, and Fabian, the mission's crew would include the pilot, Navy Commander Frederick H. Hauck, and a physician, Norman E. Thagard. This five-person crew was the largest that had gone up in a shuttle flight.

The Soviet Union had sent a woman, Valentina Tereshkova, into orbit in 1963, but Ride would be the first American woman to go into space. She therefore faced tremendous publicity. From the time her name was announced until well after her shuttle flight, she had to deal with hordes of reporters representing a nation driven wild by "Sallymania."

Like Amelia Earhart and other famous women pilots before her, Ride learned to put up with media and crowds. "I just flip the switch marked oblivious," she said. She did not enjoy the experience, however. She told one reporter that she found handling publicity "the hardest part about being an astronaut." At another time she said, "It may be too bad that our society isn't further along and that this [being the first American woman in space] is such a big deal."

The space shuttle *Challenger,* with Ride and the others aboard, lifted off from Cape Canaveral, Florida on the morning of June 18, 1983. During takeoff and landing, Ride acted as flight engineer, just as Bonnie Tiburzi did in her first flights with American Airlines. Ride sat behind the pilot and copilot, helping to monitor the many dials and signal lights on the shuttle's control panel. If an emergency had required it, she could have piloted the shuttle herself.

Once in orbit, Ride and Fabian launched two communications satellites: Anik-C, from Canada, and Palapa B, from Indonesia. The satellites would carry telephone and other signals to people in northern North America and Southeast Asia, respectively. Later in the six-day mission, they used the robot arm to launch and then retrieve the Shuttle Pallet Satellite (SPAS). This space laboratory, built by West Germany,

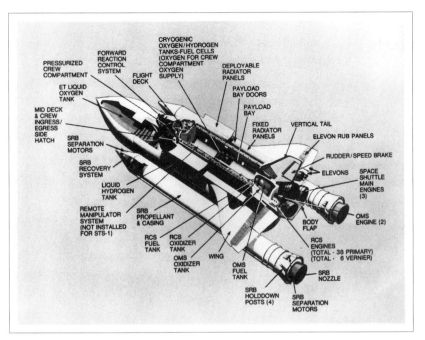

*The space shuttle, which flew for the first time in 1981, was intended to be a
"space bus"—a reusable craft that would make travel into orbit and return to
Earth relatively easy.*
(Courtesy National Aeronautics and Space Administration)

weighed over 3,000 pounds. The success of these maneuvers
showed that future shuttle astronauts could take in satellites
and repair them as well as launch them.

Ride and Fabian also activated 21 experiments, including
several tests of industrial and drug-making processes. These
processes might be carried out more efficiently in space than
on Earth. Some of the experiments were "getaway specials,"
small items that would cost relatively little to take up in the
shuttle. One getaway special was a colony of carpenter ants
that had been sent up by an inner-city high school class in
Camden, New Jersey. The ants were filmed to see whether
weightlessness changed their behavior.

Challenger landed at Edwards Air Force Base in California's Mojave Desert on June 24, 147 hours after it had left Florida. After the landing, Ride told reporters that the flight was "the most fun that I'll ever have in my life." Earlier, speaking from space, she had compared it to an "E ticket" ride, the most elaborate kind, at Disneyland.

Ride flew on another mission, the shuttle program's 13th, in October 1984. She was scheduled to go on a third flight as well. But on January 28, 1986, the *Challenger* exploded less than a minute after its launch. The seven people aboard, including two women, were killed instantly. All flight plans for the program's three remaining shuttles were cancelled while the accident was investigated. Ride was part of the investigation committee.

NASA's administrator, James C. Fletcher, also asked Ride to head a committee to evaluate the space agency's current programs and make recommendations for its future direction. On August 12, 1987, after 11 months of study, the committee turned in a 63-page report. The report was called *Leadership and America's Future in Space*. It criticized NASA's lack of direction, saying that "without an eye to the future, we flounder in the present."

The Ride report, as it came to be known, evaluated four future programs for NASA. One was a "mission to Planet Earth," in which integrated studies of Earth's land, oceans, and atmosphere would be made from satellite platforms. The other programs were sending new unmanned probes into the solar system, building an outpost for long- term living on the Moon, and making a manned trip to Mars.

Of these four possibilities, Ride most strongly recommended the "mission to Planet Earth." This project, she wrote, would allow Earth to be studied as a single system for the first time. It would show "a recognition of our responsibility to our home planet." Ride's committee also approved of launching the solar system probes. Of the other two projects, they preferred a return to the Moon over a "sprint" to Mars. "Settling Mars should be our eventual goal, but it should not be our next goal," Ride's report declared.

Ride left NASA after turning in her committee's report. She moved back to California and joined Stanford's Center for International Security and Arms Control. Two years later, she transferred to the University of California at San Diego. She presently teaches physics there. She also heads the California Space Institute, a research institute of the university that is devoted to space projects.

Sally Ride has received several awards for her pioneering work in space, including the Jefferson Award for Public Service and two National Space Flight Medals. Space exploration remains important to her. She told an interviewer that she hopes to see the United States "gain routine access to space, make it an extension of the Earth's environment, not just for exploration and national pride but for things the country can actually use—medicines, improved materials." Ride is also proud to have set an example for other women who want to go into space. She has said, "When I go out and give talks at schools, and an eight-year-old girl in the audience raises her hand to ask me what she needs to do to become an astro- naut, . . . it's neat! Because now there really is a way. Now it's possible!"

Chronology

May 26, 1951	Sally Ride born in Los Angeles, California
1968	graduates from Westlake School for Girls; enters Swarthmore College
1970	transfers to Stanford University
1973	graduates from Stanford with honors in physics and English
1977	sees NASA advertisement and applies to become space shuttle mission specialist; undergoes testing at Johnson Space Center
January 1978	accepted as astronaut trainee
summer 1978	begins astronaut training
spring 1982	chosen as crew member for seventh shuttle mission
July 24, 1982	marries fellow astronaut Steven Hawley
June 18, 1983	becomes first American woman in space
June 24, 1983	completes shuttle mission
October 1984	flies on second shuttle mission
January 28, 1986	space shuttle *Challenger* explodes
August 12, 1987	Ride turns in report evaluating future projects for NASA; leaves NASA and goes to Stanford University
1989	goes to University of California at San Diego

Further Reading

Baroni, Diane. "Flying High with Sally Ride." *Cosmopolitan,* September 1984. Interview with Ride written over a year after her shuttle mission.

Blacknall, Carolyn. *Sally Ride: America's First Woman in Space. Minneapolis, Minn: Dillon Press, 1984. Biography of Ride for young adults.*

"Challenger's Happy Landing." *Newsweek,* July 4, 1983. Account of Ride's shuttle mission, written just after mission was completed.

"Ride, Sally K(risten)." *Current Biography* 1983. Describes Ride's life through the time of her first shuttle flight.

Ride, Sally, with Susan Okie. *To Space and Back.* New York: Lothrop, Lee & Shepard, 1986. For young adults. Describes life aboard the space shuttle; illustrated with many photographs.

———, with Tom O'Shaughnessy. *Voyager.* New York: Crown, 1992. For young adults. Describes unmanned space probes that explored the solar system.

"Sally Ride: Ready for Liftoff." *Newsweek,* June 13, 1983. Provides background on Ride and describes her upcoming shuttle mission.

Sanborn, Sara. "Sally Ride, Astronaut: The World Is Watching." Ms., January 1983. Extensive profile of Ride, written a few months before her shuttle mission.

Sherr, Lynn. "A Mission to Planet Earth." *Ms.,* July-August 1987. In this interview, Ride describes the "Mission to Planet

Earth" that her committee recommended to NASA and explains why it is important.

Tyson, Molly. "Women in Outer Space." *Womansports,* February 1978. Profile of Ride written by her college roommate.

Jeana L. Yeager
(1952–   )

*In 1986, with Dick Rutan, Jeana Yeager copiloted
an experimental aircraft called* Voyager *around the
world. This was the first time an aircraft flew
around the world without stopping or refueling.*
(Courtesy The Ninety-Nines, Inc., International
Women Pilots Resource Center, Archives Department,
Oklahoma City, Okla.)

*T*he strange-looking aircraft called *Voyager* had been in the air
for three days. At his own insistence, Dick Rutan had stayed in
the pilot's seat for that entire time without break or sleep. He
knew that his copilot, Jeana Yeager, was competent, but he was

the more experienced pilot of the two. He did not trust Yeager to fly the plane while it was so heavily loaded with the fuel it would need to fly around the world nonstop. The weight of the fuel made *Voyager* unstable.

Now, however, Rutan was too exhausted to fly safely. Over the radio, *Voyager*'s ground crew begged him to let Yeager take the controls while he got some sleep. Rutan finally consented. When Yeager quietly informed the ground crew that she was in charge, one crew member said, "There's no question that at this point the more capable pilot is flying."

Jeana Yeager flew *Voyager* only about 15 percent of the time during the plane's record-setting 10-day journey in December 1986. The flight, however, would have been impossible without her. Calm under stress, she soothed the more emotional Rutan, who often drove himself to the point of collapse. "Dick [flew] the airplane, and I . . . [flew] him," Yeager said later. Rutan and Yeager had their personal disagreements, but in the air they were a perfect team. "We were really one pilot," Yeager has said. "We became an extension of one another, complementing each other."

———

Jeana Yeager was born in 1952 in Fort Worth, Texas. Her parents were Francis and Lee Yeager. She is not related to Jackie Cochran's friend, famed test pilot Chuck Yeager.

Like Jerrie Cobb, Jeana loved horses before she loved planes. Indeed, she says her first spoken sentence was, "I want a horse." When she was not riding, she was running; she raced with the track team in high school. Running, she says, gave her "a feeling of sharing the beauty and strength of horses and the ease with which they flew across the land."

In 1972, Yeager married sheriff's deputy Jon Farrar, but the marriage broke up after five years. Yeager then moved to California. She worked there as a drafter (technical artist) and surveyor. She also tried skydiving a few times and then became interested in helicopters. She was told to get a fixed-wing flying license before trying to fly a helicopter. She got her private pilot's license in 1978.

Yeager even had a brief brush with space travel. During the late 1970s, she worked as a drafter and designer for U.S. Navy captain Robert Truax. Truax was trying to build his own manned rocket. The rocket never flew with a human passenger, but Yeager said, "If [he] had offered me the chance to be the first female astronaut, I would have said yes."

In 1980, at an air show in Chino, California, Yeager saw a display of recreational planes designed by Burt Rutan. Rutan's popular planes, such as the Vari Eze and Long EZ, were sold as kits. Customers bought the designs and the parts and built the planes themselves.

At the display, Yeager spoke with Dick Rutan, Burt's older brother. Dick was as skilled a pilot as Burt was a designer. He had flown 395 combat missions in the Vietnam War and reached the rank of lieutenant colonel in the air force. He had been awarded five Distinguished Flying Crosses, 16 Air Medals, a Silver Star, and a Purple Heart.

Yeager and Dick Rutan were attracted to each other immediately. Rutan wrote later that Yeager was "all I could have dreamed of in a woman, beautiful, smart—and a pilot to boot." By the end of 1980, she had joined him in the small desert town of Mojave, California. Rutan worked for his brother at the time, but he and Yeager began talking about starting their own aircraft company.

Early in 1981, Yeager and the Rutan brothers were discussing Dick's ideas for his company over steak teriyaki at a local restaurant. Dick wanted publicity to attract attention to the company. Burt suggested they build a plane that Dick and Jeana could fly around the world without landing or refueling. Such a journey had never been made; it was the "last first" in aviation. (By refueling in the air, planes had flown around the world without landing, but they had never made such a trip without refueling.) The plane would have to be unbelievably strong and lightweight to fly with the heavy load of gas it would need to carry. Burt began sketching a possible design on his restaurant napkin.

Jeana Yeager listened intently to Burt's plan. Then she said, "Why not? Let's do it." Later she used her drafting skills to expand Burt's designs into formal plans. She also gave the aircraft its name: *Voyager*.

The three knew they would need many kinds of help to make *Voyager* a reality. They persuaded several corporations to provide materials and equipment for the project. For example, King Radio donated all of *Voyager*'s electronic gear. Companies called Hercules and Hexcel provided the materials of which the plane would be made. Teledyne Continental donated two engines. Several other businesses also helped.

Few corporations, however, were willing to give money to the project. Yeager suspected that most of *Voyager*'s cash would come from individuals' donations. She started the *Voyager* Impressive People (VIP) Club, which people could join by contributing $100. Those who could not afford that much sent smaller amounts. One man mailed in two dollars with a letter saying, "Don't laugh; I didn't get lunch today."

Individuals also donated most of the labor needed to build the plane. Helpers included family members, friends, and people who simply stopped by to see what was going on in Hangar 77 at the Mojave airport. Many stayed on, fascinated, to join the project. Some were experts in weather prediction, aerospace engineering, or other technical fields. Volunteers put in 22,000 person-hours during the two years needed to build the plane.

The building of *Voyager* began in spring 1982. By then, Burt Rutan had worked out a design quite different from the one he had drawn on the restaurant napkin. Dick Rutan later wrote that *Voyager* "looked like a catamaran sailboat crossed with a glider." The strange-looking craft had slender wings 111 feet long—longer than the wings on one of Bonnie Tiburzi's Boeing 727 jets. A boom, or outrigger, parallel to the fuselage (the main body of the plane) was placed about halfway along each wing. Near the nose of the plane was a smaller winglike structure called a canard, which braced the outriggers and gave the plane extra lifting surface. The fuselage had no tail, but a vertical fin rose from the rear of each outrigger. The fins helped to stabilize the plane as a tail would have. The plane had one engine in the front and another at the rear. The rear engine would do most of the work; the front one would be used mainly during takeoff and landing. Yeager and Rutan, as pilots, would share a space in the fuselage only seven and a half feet long, two feet wide, and two feet high—a "horizontal telephone booth." Most of the rest of the plane would be filled with 1,489 gallons of fuel in 16 tanks.

In a sense, *Voyager* was made mainly of cloth and paper. The cloth, made by Hercules and called Magnamite, was woven

Voyager, *designed by Burt Rutan and made of space-age materials that were lightweight but very strong, "looked like a catamaran sailboat crossed with a glider." At takeoff, almost all of the plane's hollow structure was filled with gasoline.*
(Courtesy Voyager Aircraft, Inc.)

from graphite fiber. (Graphite, a form of the element carbon, is also found in pencil leads.) *Voyager*'s volunteers had to make molds, place layers of Magnamite inside them, and then bake them to make the finished parts. Ninety percent of the plane would be formed from this substance, which was very lightweight yet stronger than steel. Most of the rest would be a special paper called Hexcel. The paper was soaked in resin and then formed into a honeycomb of hollow, six-sided cells. Placed between two layers of Magnamite, the Hexcel acted as insulation.

Voyager flew for the first time on June 22, 1984. In this and later test flights, Rutan and Yeager found the craft hard to control. It pitched up and down like a small boat in a stormy sea. They also discovered that moving around in the tiny cockpit required the agility of an acrobat.

Many more problems surfaced during the next two and a half years. The project's inventive experts fixed most of them, but some defied solution. One of the worst was "pitch porpoising." When the plane flew heavily loaded with fuel, the tips of its long, flexible wings sometimes bent upward. That forced the fuselage down. This in turn made the nose and canard rise. The series of movements repeated itself, making the plane fly with a wavelike motion like a swimming dolphin or porpoise. Worse still, the movements grew greater each time they occurred. If the pilot did not control them within seconds, they could tear the plane apart.

Dick Rutan believed that only he had the piloting skill to control the porpoising. He therefore never let Yeager fly the plane when it was above the weight where porpoising could occur. This meant he would have to be the pilot during the entire first three days of *Voyager*'s round-the-world flight, when the plane was heaviest. Yeager, however, felt she could learn to handle the porpoising if Rutan would let her try.

This disagreement was typical of the personal friction that developed between Rutan and Yeager as they flew more often together. Yeager recognized that Rutan was the more experienced and perhaps more talented pilot. "He shared a oneness with his airplane the way I did with my horses," she wrote later.

But she also felt that he did not respect her skill enough or give her enough opportunity to fly *Voyager*. "Dick . . . bullied away much of my confidence." Rutan, for his part, thought Yeager did not spend enough time gaining experience in other planes. She did, however, qualify for a commercial pilot's license with ratings for instrument flying and multiengine planes before *Voyager*'s long trip began.

The tension increased as more problems appeared in the aircraft. Both Rutan and Yeager began to feel that *Voyager* was "a fundamentally unsafe craft, a flying death-trap." Rutan expressed his frustration by threatening to quit the project or by shouting at Yeager. Yeager, in contrast, grew more calm and determined with each setback. When they were not working together, however, she withdrew from Rutan. She stayed in the project's office, which she managed, or went out to ride her horse, Gem.

One of *Voyager*'s volunteers had told the team two old sayings of pilots: "The airplane will never be totally ready" and "The weather will never be as good as you want." Nonetheless, on December 13, 1986, everyone agreed that plane, weather, and pilots were as ready for the round-the-world flight as they could get. That afternoon Rutan flew *Voyager* to nearby Edwards Air Force Base. The base, built on a dry lake bed, had been the scene of Jackie Cochran's breaking the sound barrier and Sally Ride's space shuttle landing. Edwards had a 15,000-foot-long runway, the world's longest. *Voyager* was likely to need every inch of it to get off the ground with a full load of gas.

Back in a trailer at the Mojave airport, the project's ground crew also made final preparations. The trailer was crowded with radios and other electronic gear. Some of this would let the crew receive weather information from space satellites. Some would let them communicate with Rutan and Yeager in *Voyager*.

Voyager took off just after eight o'clock the next morning. It became airborne with only 1,000 feet of the runway to spare. The dreaded pitch porpoising did not start, but the plane's wingtips flexed downward so much that they dragged on the

Adventures on Voyager's round-the-world flight, the route of which is shown here, included failures of key equipment and an encounter with a major tropical storm.

(Courtesy Katherine MacFarlane)

ground. The dragging tore off the winglets, small upright airfoils on the tips of the wings. At first, Rutan and Yeager were afraid the wing gas tanks might have ruptured too, but a chase plane assured them that they were not leaking fuel.

In spite of the flight's rough beginning, the two pilots were happy to be on their way at last. Yeager wrote later, "the cockpit was filled with . . . strong silence."

The project's weather expert had recommended that *Voyager* fly on the edges of the tropical storms that dotted the Pacific Ocean. This would let the plane pick up tailwinds from the storms. The tailwinds would push the craft forward, giving it extra speed without requiring more fuel. But the maneuver was risky: the wind and bumpy air inside a storm could easily destroy the fragile craft.

Dick Rutan, determinedly remaining at the controls, skirted many small storms on the way across the Pacific. Near the end of the journey's second day, however, as *Voyager* approached the island of Guam, Rutan faced a full-sized typhoon that had been given the name Marge.

Driven by 80-mile-per-hour winds, the clouds of the giant storm swirled in a spiral. To get the best tailwinds, Rutan had to fly between two arms of the spiral. At the same time, he had to avoid both the center of the storm and a smaller storm that was approaching Marge. In doing so, he dodged towering clouds that reminded him of trees in a dense forest. He gained a tailwind that boosted *Voyager*'s speed to 147 miles per hour.

Soon afterward, Yeager and Rutan noticed that *Voyager* seemed to be using much more fuel than had been expected. They could find no sign of fuel leaks, and the plane flew as if the weight of most of the "missing" fuel was still there. But Yeager had been keeping careful track of fuel consumption, and she was sure her figures were accurate. If extra fuel really was being used, *Voyager* would not have enough gas to get all the way around the world.

The two pilots discovered the answer to the fuel mystery as they were flying over eastern Africa. Fuel, they found, sometimes flowed backward during transfers from the plane's storage tanks to the tank that fed the engines. Yeager's fuel gauges did not register this, so they showed more gas being transferred and used than was really the case. (Yeager had seen and mentioned similar occurrences during test flights, but Rutan and the others had ignored her.) Recalculations now suggested that *Voyager* probably did have enough fuel to complete its flight.

Rutan and Yeager had planned to fly over southern Africa, but their radar and the ground crew's weather reports showed that area covered by storms. To avoid the worst storms, Rutan decided to risk flying over several countries in central Africa that were engaged in civil wars. The pilots had been warned to stay away from these countries because the countries' air forces might think they were enemy aircraft and try to shoot them down. Rutan, however, thought the weather looked more dangerous than military missiles. He proved to be right; the pair's narrowest escape in Africa came when they almost flew into a mountain.

When Yeager and Rutan saw the western coast of Africa pass below them, they felt such a mixture of exhaustion and joy that tears streamed down their faces. "We've made Africa, and we're on our way home," Rutan radioed to ground control.

But the trip was far from over. On the seventh day, just north of the coast of Brazil, Rutan and Yeager found themselves in the middle of a thunderstorm—at night. *Voyager* went completely out of control, slowly turning upside down. As Rutan put it, "The weather was flying the airplane." When the storm finally spat them out, the plane was almost completely on its side. To get it level again, Rutan had to send it into the steepest dive it had ever made. His plan worked, but he recalled later, "I had never felt this unsafe before, not [even] in combat." A little later, Yeager had to take over the controls because exhaustion made Rutan "hit the wall," as military pilots say. Rutan said later, "I couldn't remember how to do anything . . . and I didn't care."

The journey was nearing its end—but every time Yeager and Rutan relaxed, something else happened. On the last day, they began to worry again that they might not have enough gas to complete the trip. They started draining all their remaining fuel into the tank that fed the engine. But one fuel pump failed to work right, and air instead of gas got into the line. Without fuel, *Voyager*'s rear engine, the only one running, died. The plane went into a dive.

With Yeager calmly reading the checklist of procedures, Rutan tried to restart both the front and the rear engines. After long, heart-stopping moments, the front engine caught. Rutan used it to level out the plane. Then fuel from another tank could flow to the rear engine, and it, too, began running again. Once the engines were stable, Rutan was able to replace the defective pump.

Voyager landed at Edwards Air Force Base in the early morning of December 23, 1986—a full day ahead of schedule. Rutan and Yeager had flown 25,012 miles in nine days, 3 hours, and 44 minutes. Huge crowds welcomed them home.

On December 29, President Ronald Reagan awarded Rutan and Yeager the Presidential Citizens Medal, which is given to Americans who "have performed exemplary deeds of service for their country or their fellow citizens." Reagan said that Rutan and Yeager were "living examples of American pioneerism at its best." The Smithsonian Institution gave them the National Air and Space Museum Trophy shortly afterward.

Voyager joined other historic aircraft in the National Air and Space Museum's permanent display. In addition to being a heroic feat of endurance, its flight had demonstrated the usefulness of lightweight composite materials and some of Burt Rutan's design features. These things may be used in future scientific or military planes that need to remain in the air for a long time. As one reporter wrote, *Voyager* "made us all believe in paper airplanes."

Rutan and Yeager continued to make public appearances together, but in private they lived apart. Their personal relationship had really ended long before the flight; only the all-consuming dream of *Voyager* had kept them close. As Yeager said, "Most people break up and go their separate ways. We broke up and stayed together."

Near the end of the 1980s, Yeager did go her separate way. She fulfilled her long-postponed dream of learning to fly helicopters. She also took up harness racing. She is now married to Bill Williams, the inventor of a chemical that reduces corrosion in metal aircraft parts. Williams admiringly calls Yeager his "95-pound bulldozer" because of her determination. The two live in Bellingham, Washington.

Dick Rutan and Jeana Yeager were often asked whether *Voyager*'s flight was worth all the effort and heartbreak it cost them. In the book that described the flight, Yeager gave her answer: "Yes, it was . . . worth all the work and sweat and patience, worth the physical discomfort and the danger and the fear. . . . And it would have been worth the effort even if we had failed or lost the *Voyager,* or lost our lives."

Like all of aviation's classic flights, by both men and women, *Voyager*'s flight showed what individuals could do when they were willing to risk everything to achieve their dreams. In doing so, they extended the bounds, not only of their own lives, but of the human race. In the deepest sense, they captured a piece of the sky.

Chronology

1952	Jeana Yeager born in Fort Worth, Texas
1977	moves to California
1978	earns private pilot's license
1980	meets Dick Rutan; moves to Mojave, California
1981	with Rutan brothers, begins plans for flying around the world
Spring 1982	building of *Voyager* begins
June 22, 1984	first test flight of *Voyager*
December 14, 1986	*Voyager* begins round-the-world flight
December 16, 1986	flies close to Typhoon Marge
December 19, 1986	cause of missing fuel discovered
December 19–20, 1986	*Voyager* flies across central Africa
December 21, 1986	storm sends plane out of control
December 22, 1986	engine quits
December 23, 1986	*Voyager* lands after flying around the world
December 29, 1986	Yeager and Rutan awarded Presidential Citizens Medal
1987	Yeager and Rutan awarded National Air and Space Museum Trophy
late 1980s	Yeager leaves Voyager, Inc.; marries Bill Williams

Further Reading

"Around the World in 11 Days." *Newsweek,* September 22, 1986. Describes plans for *Voyager* flight a few months before it took place.

Garrison, Peter. "*Voyager* Flight Fantastic." *Flying,* March 1987. Account of *Voyager's* flight.

Moses, Sam. "Gallant Victory for an Odd Bird." *Sports Illustrated,* January 5, 1987. Account of *Voyager's* flight.

Park, Edwards. "The *Voyager's* Bid to Girdle the Globe Is No Mere Canard." *Smithsonian,* February 1985. Describes plans for *Voyager's* flight almost two years before it took place; provides background on Rutan brothers and Yeager.

"Up, Up and Around." *Newsweek,* December 26, 1986. Account of *Voyager's* flight and review of background on the plane and its pilots.

Yeager, Jeana, and Dick Rutan, with Phil Patton. *Voyager.* New York: Knopf, 1987. Story of the building, testing, and round-the-world flight of *Voyager.*

"Yeager, Jeana." *Current Biography,* 1977. Provides background on Yeager's life and *Voyager's* flight.

INDEX

Bold numbers indicate main headings.
Italic numbers indicate illustrations.

Index

Index